THE
PLANETS

THE
PLANETS

David McNab and James Younger

Yale University Press
New Haven and London

This book has been published to accompany the BBC/Arts and Entertainment television series *The Planets*.

Published by BBC Worldwide Ltd, Woodlands, 80 Wood Lane, London W12 0TT
Published in the United States by Yale University Press

Executive Producer: **John Lynch**
Series Producer: **David McNab**
Producers: **James Younger, Jacqueline Smith and Becky Jones**

ISBN 0-300-08044-1

Library of Congress Cataloging-in-Publication Number: 98-89995

A catalogue record for this book is available from the British Library.

Commissioning Editor: **Sheila Ableman**
Project Editor: **Lara Speicher**
Consultant: **David Hawksett**
Copy Editor: **Patricia Burgess**
Art Director: **Linda Blakemore**
Designer: **Bobby Birchall, DW Design**
Picture Researchers: **Miriam Hyman, Chris Riley, Sean O'Curneen**

Set in Cosmos and Frutiger by DW Design
Printed in Great Britain by Butler & Tanner Ltd, Frome and London
Color separations by Radstock Reproductions Ltd, Midsomer Norton
Jacket printed by Lawrence Allen Ltd, Weston-super-Mare

The paper in this book meets the guidelines for permanence and durability of the Committee on Production Guidelines for Book Longevity of the Council on Library Resources.

10 9 8 7 6 5 4 3 2 1

Acknowledgements

The authors want to thank the team that made *The Planets* television series possible. They are: Dr Chris Riley, who enthused us with the sheer wonder of the Solar System and awakened in us both an unexpected interest in asteroids; Jacqueline Smith, who seemed able to tackle the entire Russian space industry single-handedly; Sean O'Curneen, whose explanations of complex astrophysics both illuminated and baffled us; Angela Pierce and Gina Seddon who provided organisational genius and support. We are also indebted to our Executive Producer John Lynch for his kind wisdom and encouragement, and to Lara Speicher from BBC Books for her guidance and unending patience.

From the world of planetary science we would like to acknowledge the valuable help of Hal Levison, Carolyn Porco, Bruce Murray, Brad Smith, Jim Head, Sasha Basilevsky, Alvin Seiff, Imre Friedmann, Steve Mojzsis and David Grinspoon.

Finally, James couldn't omit to thank his mother and father – for getting him this far – and Susie, who endured a year of lost weekends but still managed to be a bright shining star in his skies.

David thanks his mother, Rita, and the in-laws, Robin and Lesley Ward, for helping to keep the world turning while he was away in a darkened room. And, lastly, to the adorable Sharon, Millie and Nancy for still being around when he had finished.

contents

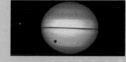
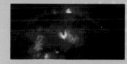

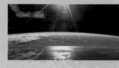

preface

THE YEAR 1959 is barely two days old when mission controllers at the Baikonour Cosmodrome start the countdown. Outside, a small army of scientists clusters together in tiny pockets, chatting nervously, their frigid breaths mingling against the night sky. They are braving a ferocious winter deep in the wilderness of Kazakhstan to witness one of the most breathtaking sights of the age: the firing of a Soviet *SL3* launcher. Suddenly the talking stops, the ground shakes and they feel the growing rumble of a rocket stirring into life. Bright clouds billow on the horizon, and inside launch control, above the hubbub of countless running checks, a single calm voice recites a list of descending numbers. 'Tri…dva…raz…pusk!' The noise ratchets up a notch and the 281-tonne shaft of metal rises impossibly slowly from the ground. As the sound fades and the accelerating fireball disappears behind a cover of low cloud, there's a sense of cautious excitement in the control room. Minutes later the tiny spacecraft, *Luna 1*, nicknamed *Lunik* (Little Moon), and the final stage of the rocket are orbiting the Earth, waiting for the momentous command. Then the booster gives the craft one last almighty shove. *Lunik* speeds towards the Moon and passes a milestone – it is the first craft to head out into space.

Lunik 1 was not a successful mission in a normal sense. Within hours of the launch, Soviet engineers realized that their spacecraft was off course. Two days later it flew wide of the Moon by some 6,000 kilometres. But as the downcast flight controllers mulled over the fate of their wayward craft, an inspired thought occurred to them. *Lunik 1* might have failed in its primary objective, but in doing so it had inadvertently sailed into history: it had become the first man-made object to join the nine known worlds that are caught in the gravitational embrace of our Sun. Suddenly, 'Little Moon' seemed inappropriate for a craft embarking on an endless journey around the Solar System. It was now in the company of the planets and the Soviet scientists renamed it *Mechta* – The Dream. The dream for scientists in the new field of space exploration was that this tenth 'planet' would soon be joined by others – probes that would visit the other worlds in the Solar System and send back news of their alien landscapes.

The 20th century might be remembered for many things: the furry growth of a penicillin mould in a glass dish; the discovery of a genetic code written on a molecule of DNA; the forging of logic in a microscopic sliver of silicon; the unleashing of the immense power inside the atom… It is too soon to guess what value our descendants will attach to these achievements. But the hundred years that separate the end of the age of steam from the dawn of the third millennium are the proud owners of one clear fact: this was the age when we first broke free of our planet and ventured out to other worlds orbiting the Sun.

This book chronicles mankind's voyage to the planets as it stands at the end of the first 40 years of space travel. For 4,000 years the Egyptians, the Chinese, the Greeks, the Arabs and the Europeans had gazed at bright points of light in the night sky and stumbled towards an understanding of our corner of the Universe. The Earth, they had found, was a planet revolving around the Sun. The five wandering stars were other planets, locked in a similar dance around our central star. Astronomers had trained their telescopes on them, measured their size and their speed, and even added three more planets unknown to the Ancients to the family of the Solar System. But the planets remained faceless spots of light for the most part, fuzzy discs at best. Just as nature deplores a vacuum, so the human mind detests lack of detail and imaginations ran riot. Early astronomers imbued the planets with landscapes, oceans, wildlife and civilizations. The sky was filled with alien, alternative Earths.

Three decades after *Mechta*, America's unmanned probe, *Voyager 2*, floated past Neptune and brought the first stage of planetary exploration to a glorious end. In those 30 short years, human beings have sent robotic emissaries to every planet apart from Pluto, discovered dozens of new moons in orbit around other planets, and put to rest the myths and fantasies that had been accepted for centuries. But the space age replaced fantasy with even more fantastic fact. Volcanoes three times the size of Mount Everest, storms twice the size of the Earth, worlds with seas of methane, rivers of lava longer than the Nile, clouds of sulphuric acid, frosts of pure shining metal: these were the sights awaiting us out in the realm of the planets. This is the story of those exhilarating discoveries.

LEFT *A dream is born: the age of space exploration began in earnest on 2 January 1959, when* Luna 1 *blasted off for the Moon.*

different worlds

FROM OUT OF NOWHERE, through the swirling fog, a world the size of the Earth's Moon looms into view. Its surface is glowing with lakes of bubbling lava, and a torrent of impacting meteorites sends white-hot rock spraying high up into its wispy jacket of sulphur-green clouds. The bombardment is relentless and, stone by stone, rock by rock, boulder by boulder, the growing planet absorbs anything that dares stray into its path. Originally just one of a huge army of rocks circling an embryonic Sun, this world has grown large. It has swallowed up countless smaller rivals. Others, which narrowly escaped its draw, it threw off course and sent to a fiery death in the heart of the Sun. The odds of this young planet surviving are slight — at any

moment it could be torn apart in a cataclysmic collision with another large ball of rock. There are scores of others like it in this young Solar System, and only a handful will make it. Yet somehow it does survive. In 100 million years it will grow large enough to hang on to a thicker layer of gas and cloud. Its surface, now hot enough to melt rock, will cool to a temperature where water will condense into oceans and, one day, it will develop life. In 4.6 billion years' time a late-blooming species will call it the Earth. Their small robots, honed from remnants of the exotic metals that now rain down from the heavens, will rocket out of this planet's gravitational pull and venture out into the void to visit the other survivors of this battle zone. Bit by bit, the little spacecraft will send back news of these veterans, and the story of the creation of the planets will unfold.

Clyde Tombaugh arrived at the Lowell Observatory in Flagstaff, Arizona, on 15 January 1929. A Kansas farm-boy turned amateur astronomer, he was itching to get his professional career under way. He brought with him a suitcase and not much else, not even the money for a return ticket home. Tombaugh was greeted at the station by the observatory director, Vesto Slipher, and given a brief tour of the institution that would be his home for the next few years. On the way to his quarters, Slipher introduced him to a 13-inch photographic survey telescope still in the process of being built. Over the next year, the young man was going to get to know every nut, bolt, cog and gear on that telescope. With its help he would discover a planet and define the limits of the Solar System.

PAGES 8–9 *When two worlds collide: this computer-generated image captures the violence that was routine in the early days of the Solar System.*

Although it probably would not have occurred to him at the time, Tombaugh was joining the ranks of an ancient profession, the latest in a line of stargazers dating back more than three centuries. In 1609 a Florentine genius called Galileo Galilei had become the first human being ever to point a telescope towards the night sky. He had found that the stars, even magnified many times, could still only be seen as pinpoints of light. But to his delight Galileo also saw a different breed of object up there, which appeared not as dots but as tiny bright discs against the blackness of space. He was seeing up close the objects the Ancient Greeks had called *planetos*, or 'wanderers'. The planets were not fixed in constellations, like the stars, but from night to night, month to month, they could be seen slowly changing their positions in the firmament.

ABOVE LEFT *The recently restored Pluto dome at the Lowell Observatory in Flagstaff, Arizona, houses the photographic telescope with which Clyde Tombaugh captured Pluto in February 1930.*

ABOVE RIGHT *Portrait of a planet hunter: Clyde Tombaugh poses with a wooden photographic-plate holder. One of those plates would prove to hold – somewhere among hundreds of thousands of stars – the image of the last planet.*

Apart from the Sun and the Moon, the Ancients saw five different bodies wandering through the heavens. So important did they seem that they named them after their gods. The bright one gently sweeping across the night sky was named Marduck by the Babylonians, Odin by the Norse, and Zeus by the Greeks. It was the Romans who called it Jupiter. The faint, fast-moving point of light, never far from the Sun, the Romans named Mercury, after the messenger of the gods. The most brilliant they named Venus, after the goddess of love and beauty; the blood-red one was Mars, in honour of the god of war; and the slowly drifting one was named Saturn, after the god of time. At the time Galileo spied them through his telescope, an idea that is universally accepted today was just taking hold: these wandering discs were worlds just like

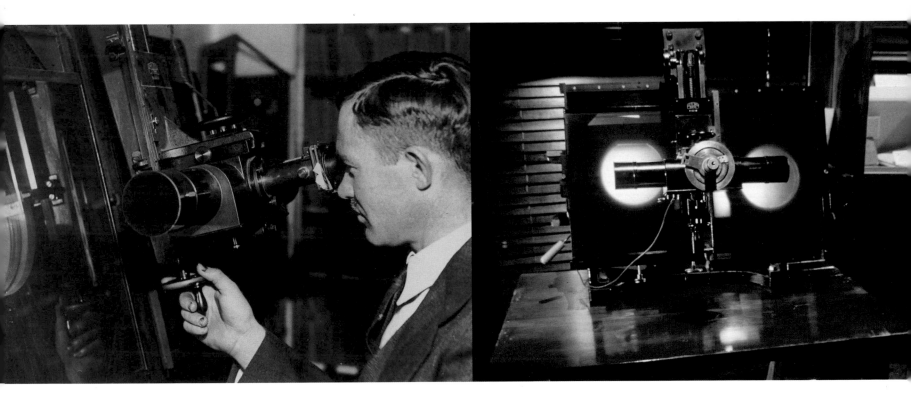

the Earth. To Galileo the Solar System was a family of six planets orbiting the Sun. Mercury and Venus were the closest in; out beyond the Earth lay Mars, Jupiter and finally Saturn.

By the time Clyde Tombaugh began working in Flagstaff, there were already two additions to the family. In 1781 a telescope many times more powerful than Galileo's had spotted Uranus, moving in an orbit twice as far away as Saturn. Then, in 1846, came the sighting of Neptune, hiding in the dark fringes of sunlight, 30 times as far from the Sun as the Earth. Neptune marked the end of the known Solar System and was the reason Tombaugh had come to the observatory. Soon after its discovery, astronomers had thought the planet seemed to be lurching from its expected path around the Sun. The possibility that the gravitational pull of an even more distant world might be tugging Neptune off course had instigated a search for a new, even more distant planet – code-named Planet X. The search had been going on for 40 years when the boy from Kansas came on to the scene.

Tombaugh had no formal training in astronomy; he taught himself about the planets from a book that his uncle had lent him. During his teenage years he built several telescopes out of bits and pieces from around his parents' farm and would spend every spare moment observing the sky and sketching the planets. He was 22 when his father's crop was ruined and he determined to look for work to help his family through the coming winter. He wrote to the Lowell Observatory because it was the only one he'd heard of and to his utter surprise he got the job. By February 1929 the new telescope had been finished and Slipher had taught

BELOW LEFT *Nestling among the hundreds of photographic plates in the basement of the Lowell Observatory are Tombaugh's original images of Pluto.*

ABOVE *After several months of media attention, the deeply modest Clyde Tombaugh returned to the envelope containing the plate on which Pluto appears and wrote, 'Planet "X" (Pluto) at last found!!!'*

Tombaugh the rudiments of astronomical photography. From then onwards, Clyde Tombaugh was left alone with the telescopic camera and a sky full of stars.

Blink-blink

The principle of the search for Planet X was simple but laborious. The first stage was to take pictures of a tiny patch of stars several nights apart. The two photographic plates were then to be viewed in a blink comparator. This machine allowed the astronomer to compare the plates by flicking, or blinking, between the two supposedly identical images. Only by subjecting every single star to this scrutiny would it be possible to see if, over the course of a few days, one of the stars had wandered from its original position, signalling that it was in fact a planet.

By the summer, Tombaugh had generated a mountain of plates awaiting this painstaking search and Slipher, rather than tying down a more experienced astronomer, decided to offer the young man the opportunity to clear the backlog himself. Tombaugh was overwhelmed. He realized that the person studying the plates in the blink comparator was, in effect, the person entrusted with discovering a new world, the ninth planet.

BELOW *The discovery of Pluto – the unimaginably small speck of light that over the course of six days had shifted across the sky. Would you have seen it?*

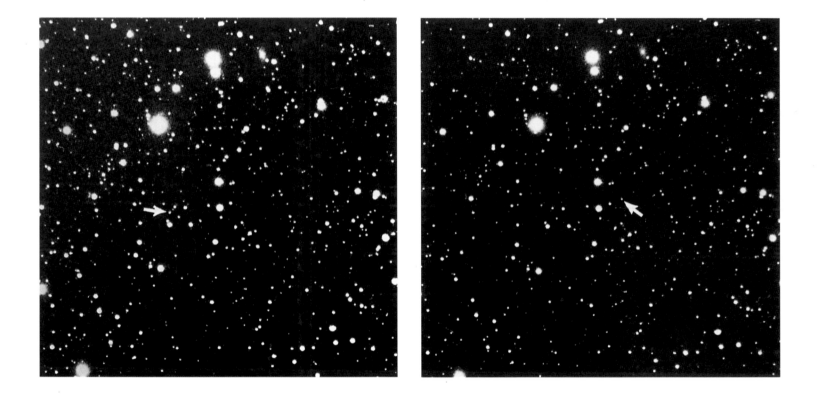

Nevertheless, no one said it would be easy. Each photographic plate could include anything from 50,000 to a million stars: it could take a week or more to complete a comparison of just one pair. Tombaugh worked in sessions of three to six hours; any longer and he would become exhausted. As if hours of intense scrutiny of sections of plates containing hundreds of thousands of stars weren't difficult enough, the young astronomer had to be continually alert to false alarms: smudges on the plates caused by drifting asteroids and comets, even adjacent stars whose brightness would inexplicably change. Then, of course, there were the known planets – the seven discs of light, varying in size and brightness from a dim speck to a glowing ball, which would appear from time to time across his plates.

By January 1930 Tombaugh's photographic plates formed a patchwork of images that together made up a 360-degree strip that circled the entire night sky. On 15 February he placed another two plates in the blink comparator. They were of a region around the star Delta Geminorum that had been taken four weeks earlier. For three days he blinked one plate against the other until he had confirmed a quarter of the points of light as stars. Then he found it. Despite tired eyes and months of false alarms, when Tombaugh saw a faint speck of light changing position as he flicked backwards and forwards between the two plates, he had no hesitation – it *was* Planet X. He later recalled, 'A terrific thrill came over me. I switched the shutter back and forth, studying the images. [I thought] I had better look at my watch and note the time. This would be an historic discovery.' Tombaugh's watch told him it was 4.00 p.m. on

18 February 1930. That night he went into Flagstaff and celebrated by watching Gary Cooper in *The Virginian*.

As soon as Vesto Slipher announced the discovery, telescopes around the world were trained on Tombaugh's planet, now called Pluto. What they saw was a surprise. Pluto didn't meet anyone's expectations. Planet X was supposed to have been a large world, capable of pulling Neptune off track. Instead, Pluto was tiny, smaller even than the Earth's Moon. It was several years after Tombaugh's discovery when astronomers realized that the apparent deviation of Neptune's orbit was an illusion – a result of it having been studied for much less than one Neptunian year. The mysterious Planet X did not exist at all. But in discovering Pluto and confirming that there were no more large planets out there waiting to be discovered, Tombaugh had set the scale of the Solar System. The family was complete.

ABOVE *Still an enigma: 64 years after its discovery, the Hubble Space Telescope took this fuzzy picture of Pluto, accompanied by its moon, Charon (right). It is the best image of these distant worlds we have to date.*

Meet the family

The Solar System's first world is about 100 times closer to the Sun than Pluto. A close study of Mercury was impossible until the space age – the planet never wanders far enough from the glare of the Sun for astronomers to be able to see it clearly. Mercury is a rocky planet just twice the size of Pluto, but what it lacks in size it makes up for with speed. It takes a mere 88 Earth days to complete its orbit of the Sun: a year on Mercury therefore lasts about three Earth months. In

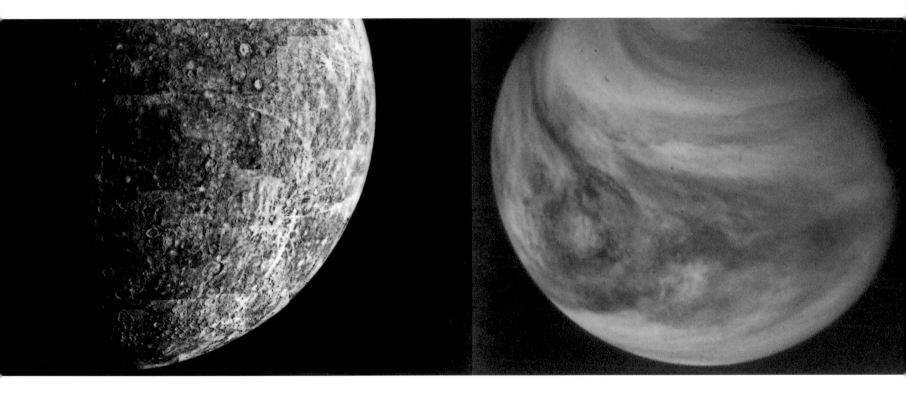

PAGES 16, 17, 18 *The family of planets as Tombaugh knew it, before his discovery of Pluto. In order of their distance from the Sun (not including Earth): Mercury, Venus, Mars, Jupiter, Saturn, Uranus and Neptune.*

1974 the American space probe *Mariner 10* sent back the first images of Mercury's cratered surface and revealed a planet that looked like a larger sister to our Moon.

Halfway between Mercury and the Earth lies Venus, the second of the four rocky worlds in the Solar System, and the brightest planet in the night sky. Like all the planets, Venus spins as it travels around the Sun. The time it takes for a planet to complete one revolution on its axis is known as its day; the Earth takes 24 hours to do this, but Venus spins incredibly slowly. A Venusian day takes nearly 243 Earth days – longer than it takes the planet to orbit the Sun – so a day on Venus lasts longer than its year. Venus was once believed to be our heavenly twin, since it is not just the closest planet to the Earth but is also nearly identical in size. But in October 1967 a robotic Soviet probe plunged beneath the cloud tops to reveal the terrible truth about our twin: beneath its bright clouds lies a searing, lava-filled hell-hole.

The next planet lies about 80 million kilometres beyond the Earth — 230 million kilometres from the Sun. Through a telescope Mars is a faint, reddish disc, half the size of the Earth. It's the last of the four rocky planets — a desert world swept by dust storms. Although it is much smaller than the Earth, it shares some close ties with our world. Mars spins once every 24 hours and 36 minutes, so a day on Mars is almost the same as ours. It also has similar seasons because of the way it is tilted. As they move around the Sun, some of the planets rotate bolt upright (with the poles at due north and south), but most of them are tilted. The Earth, for example, is tilted by some 23 degrees from the vertical and it is this tilt which is responsible for

our varying seasons as we travel around the Sun. Mars has an almost identical axial tilt, 25 degrees off vertical. Mars' much longer journey around the Sun, however, means that its year lasts 687 days.

On nights when the Moon and Venus are not visible, Jupiter is the brightest object in the sky. This massive planet, circling nearly 800 million kilometres from the Sun, could swallow all the other planets in the Solar System with room to spare. In the late 17th century, Italian astronomer Gian Domenico Cassini made out the ghostly traces of banded clouds on Jupiter and a giant storm that survives to this day — the Great Red Spot. This was mankind's first inkling of the true nature of the planets beyond Mars. Jupiter, big enough to contain 1,300 planets the size of the Earth, isn't made of rock and metal — it is a gas giant, a huge ball of swirling vapours with an atmosphere thousands of kilometres thick.

Even the feeblest telescope will reveal that Saturn has the most eye-catching adornment in the entire Solar System: a set of magnificent pale rings that circle its equator. Beneath the rings lies a planet that is a smaller cousin of Jupiter, covered with storms of cream, ochre and sepia-coloured clouds. Far beyond are the other two giant planets, much smaller than Saturn but still many times bigger than the rocky planets. Uranus is so far from the Sun – nearly 3 billion kilometres – that it takes 84 Earth years to complete a single orbit. Seen from Earth, the moons of Uranus revolve around it like the lights on a Ferris wheel. That is because Uranus spins on its side, rolling around the Sun like a barrel. Blue Neptune, 30 times as far from the Sun as the Earth, is so incredibly distant that from its cloud tops the Sun would

seem little more than a bright star. No one could live to watch a complete orbit of this twilight world, as a year on Neptune spans two human lifetimes: 165 years.

And what of Pluto? The best telescopes show it to be a solid planet made partly of rock, partly of ice, but no spacecraft has ever seen it close up. From the moment of its discovery, Pluto was a mystery, spurring astronomers to ask the question that had been in the back of their minds for decades. How did the planets all end up so different? Why are the four planets in the inner Solar System small and rocky, and the next four planets giant and gassy? And then there is Pluto which doesn't fit into either category. The story of our attempts to understand the birth of the planets spans several centuries, and is, in truth, a detective story whose last chapter has not yet been written. As the 20th century draws to a close, there are

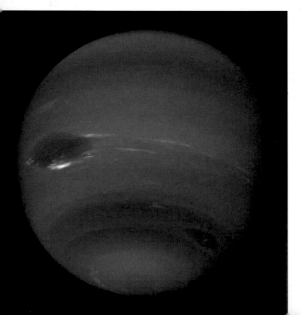

Tombaugh's
funny planet

Pluto's orbit is not a circle but a stretched-out ellipse that is tilted to the plane of every other planet. At one end of its orbit Pluto swoops high above the planets and at the other it plunges below them. Pluto is also unique in that its eccentric orbit sometimes carries it across the path of Neptune so that for several years Neptune is actually the furthest planet from the Sun. So elongated is Pluto's 248-year passage around the Sun that the difference between its closest pass and its furthest point is almost the same as the distance from Uranus to the Sun. During its summer, the frozen nitrogen on Pluto evaporates to create a temporary atmosphere. With the onset of

winter the nitrogen turns to frost and falls back to the surface. On Pluto the winter weather doesn't merely deteriorate – it completely disappears.

In 1978 Pluto was found chaperoning a companion moon, Charon, which is almost one-eighth its mass – far and away the largest moon in the Solar System in comparison to its parent planet (our Moon is only one-eightieth the mass of the Earth). Pluto and Charon lie a mere 20,000 kilometres apart – 20 times closer than the Moon is to the Earth – and take just six and a half days to spin round each other once. So close are these worlds that they are caught in a gravitational headlock, each presenting the same face to the other. From the surface of Pluto, Charon would appear as a fixed object in its sky. A Plutonian living on the opposite hemisphere wouldn't even know that Pluto had a moon.

1 *An artist's impression of Pluto and its moon Charon. NASA plans to send the first probe there in 2004. Pluto Express will take around 12 years to reach the planet if all goes to plan, and will send back the first images of Pluto and Charon.*

suggestions that Pluto, far from confusing our efforts to unravel the evolution of the planets, may turn out to be the Rosetta Stone of planetary formation – a tiny, distant clue to how these different worlds came to be. But wherever the story of the creation of the planets might end, the beginning of it, like the Solar System itself, starts with the creation of our Sun.

Four billion years BC

On 24 April 1990 the exhaust gases billowing out from launch pad 39A at Cape Canaveral in Florida signalled the take-off of space shuttle *STS-31*. Its mission was to place the Hubble Space Telescope in orbit around the Earth. Free from the atmospheric disturbances that have bedevilled Earth-based astronomers, Hubble would be able to see deep into space. In 1994 it was peering into the distant Orion Nebula when, through a thick cloud of gas and dust hundreds of billions of kilometres across, it saw an arresting sight. Bright points of light were shining through the dust – as pockets in the cloud collapsed under their own weight, stars were being born. Within this stellar nursery, Hubble found evidence that the formation of a star was not an isolated event. Stars, it seems, are born in clusters, the gravitational instability created by the birth or the death of one star causing nearby pockets of gas and dust to collapse. Stars form just as dominoes fall – one influencing the next. The spectacular Hubble images captured dramas similar to those played out more than 4 billion years ago and much closer to home when our own Solar System formed.

On the Orion arm of the Milky Way, 23,000 light-years from the centre of our galaxy, a massive cloud was gently drifting through space. At least 8 billion years earlier, the Universe and all the matter in it had been created in the event known as the Big Bang. Soon after that matter had begun to collapse and clump together to form galaxies. By now our own galaxy was already billions of years old and full of stars. As this massive cloud floated between the distant star fields, being pulled first this way then that by gentle gravitational tugs of other stars, pockets of gas in the cloud started to collapse inwards on themselves. As these imploding pockets reached a critical density, their cores heated up and ignited: stars were born. If we could watch the process of aeons speeded up into minutes, we might see small points of light flickering into life as stellar neighbourhoods sprang up throughout the cloud. Most of the smaller lights might glow steadily for several tens of seconds before slowly burning themselves out. But as the rash of stars spread throughout the cloud, a few of the lights – the largest and brightest – would flare for just a few brief seconds before disappearing. These are giant stars that burn for just a fraction of the lifetimes of their smaller, steadier siblings. When these stars die, their fate is to explode into supernovae. So violent are these supernovae that they send plumes of white-hot plasma bursting out into surrounding clouds, disrupting their uneasy equilibrium and nudging them into collapse. In one corner of that cloud on the Orion arm of the Milky Way, about 4.6 billion years ago, one of those supernovae sent a shock wave surging out at over 32 million kilometres per hour. Almost immediately a pocket of gas next to it began to contract and rotate, soon flattening out into a swirling disc of debris. Deep in the twinkling parent cloud another light flickered on – it was our Sun.

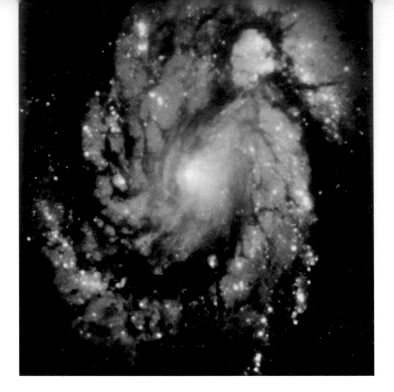

Clouds of creation

It was Immanuel Kant, the 18th-century German philosopher with something of a flair for physics, who first suggested that the planets, like the Sun, had formed from such a cloud of gas and dust. Kant had seen milky patches of light throughout the sky and had realized they were clouds of interstellar material. He also heard astronomers talk of the swirling patterns of matter dotted around the heavens – which we now know to be distant galaxies – naturally organized into flattened spirals. Kant's crude but brilliant hunch was that our planets had formed in a similar way – from a cloud of dust circling our Sun. In 1796 a popular book on celestial mechanics was published by French mathematician Pierre Simon de Laplace. In it Laplace asked a simple question: why do the planets all move around the Sun in the same direction? Surely, he argued, they must have formed from a single cloud of debris, or nebula, that once rotated about our Sun. Kant and Laplace both saw planets as the leftovers from the birth of stars. But they had no proof, and there were other ideas for the origins of planets. Some said that they had wandered free in the galaxy until they were captured by the pull of our Sun. Others suggested that a close encounter between our Sun and a passing star had ripped the planets out of the Sun.

In 1984 two planetary scientists, Brad Smith and Rich Terrile, were taking time out from their work on the *Voyager* mission to the outer planets (see Chapter 4). They had gone to an observatory high in the mountains of Chile to study the rings of Uranus and to see if they could identify any cloud movements on Neptune. Two or three months after returning to NASA's Jet Propulsion Laboratory (JPL) in Pasadena, California, Terrile finally got round to looking at a few extra images Smith had taken of a handful of young stars. Among them was one of the most startling pictures they could ever have imagined. When the star Beta Pictoris appeared on his computer screen, Terrile could barely believe what he saw: surrounding the young star was a foggy disc of dust. Perhaps the cloud was an optical aberration? 'It was a heart-stopping moment. I checked that it wasn't a fault – I checked everything but it just wouldn't go away,' he remembers. What Smith and Terrile had captured was the first image of an alien solar system in the throes of creation.

Just over 4.5 billion years ago our Solar System must have looked something like the star Beta Pictoris does today. Space would not be black; instead we would find a vast fog of spiralling dust grains and swirling gas. At the same distance from the Sun as the Earth is now there existed a comfortable room temperature; the source of this heat was a softly glowing ball of yellow light. This distant foggy glow, our young Sun, was already exerting

ABOVE *Spiral galaxy M 100 is comparable to a young solar system on a giant scale – everything revolves around a massive central core.*

FAR RIGHT *Immanuel Kant (1724–1804), the German philosopher who intuitively saw the connection between the birth of the Sun and the birth of the planets.*

BELOW *Pierre Simon de Laplace (1749–1827), the French mathematician who gave scientific credibility to Immanuel Kant's hunch about planet formation.*

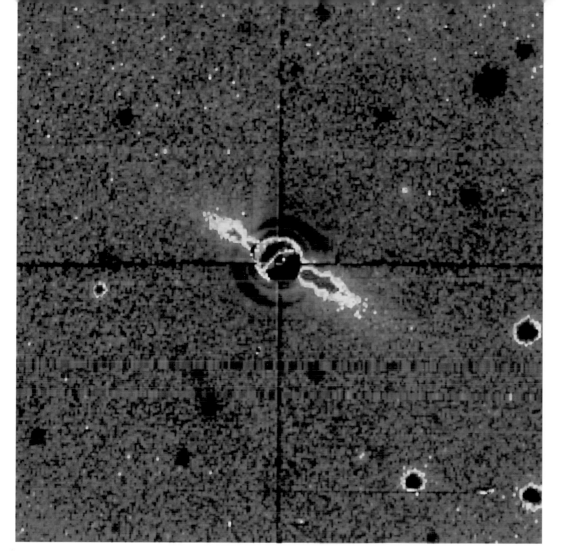

a gentle pull on the swirling pea-souper. Eventually, the growing gravitational pull from the centre would suck in nearly all the circling cloud — but not quite all; a tiny fraction of gas and dust had just enough momentum to escape its pull. This tiny amount of leftover debris, just one-fifth of 1 per cent of the total cloud swallowed up by the Sun, became the planets, became the Earth, became us. What could have happened within that remnant of gas, overlooked by our Sun, that transformed it into these nine different worlds? Halfway through the 20th century that question received an added impetus. Developments in rocketry gave us a new perspective on these worlds — soon they would be more than distant discs of light. Mankind was reaching out to the planets.

Escape from the third planet

For 14 years after the discovery of Pluto, Clyde Tombaugh continued to search the heavens. If he had found one planet so quickly, perhaps there were others. In that time Tombaugh found many things — new comets, new asteroids — but no planets. However, his reputation as a precision telescope-builder had become second to none. Then his career took an unexpected turn when he was called to the deserts of White Sands, New Mexico, to track the vertical

progress of US military rockets as they edged ever closer to space. As the prototype rockets spiralled up and beyond the scope of conventional cameras, the space pioneers needed a new means to capture their successes and, more important, their failures on film. In this way they could analyse, frame by frame, why some rockets worked and others didn't. The rockets were the brainchild of a German scientist called Wernher von Braun and had originally been commissioned by Adolf Hitler to deliver warheads to London. Some say that the space age really started in a suburb of west London, where the first *V2* rocket landed with devastating effect. But despite his wartime success, von Braun wasn't interested in weapons of destruction; he dreamt of space. When asked about the performance of his early rockets, he famously remarked that 'the rocket worked just fine – but it landed on the wrong planet'.

After the Second World War, von Braun and nearly 200 of his engineers were spirited away to the USA to develop rockets for the American army. But so advanced was his work that it brought benefits to both sides of the Iron Curtain. Having failed to acquire the German genius in person, the Soviet Union sent their top rocket engineer, Sergei Korolev, to pick over the bones of von Braun's abandoned factory. The Americans hadn't left much, but what they did leave came as a shock to the young Soviet rocket engineer: von Braun's rockets seemed impossibly advanced. Korolev returned to the Soviet Union with enough information to accelerate its rocket industry into a force to rival that of the Americans.

By 1947, with the advantage of Tombaugh's detailed telescopic films, adapted versions of von Braun's *V2*s were soaring high above our atmosphere and out into space. From their vantage point some 400 kilometres up, they photographed the one planet that Tombaugh wasn't able to see with his telescope: Earth. Within the next decade, the Soviets had launched the first artificial satellite into space. For 21 days *Sputnik* beeped its meaningless but triumphant message down to Earth: mankind had officially entered the space age. In March 1965 Cosmonaut

Alexei Leonov squeezed out through a hatch on his orbiting *Soyuz* craft. Although not the first person to be launched into space (that honour had gone to his colleague and dear friend Yuri Gagarin), Leonov was the first human being to leave the confines of a spacecraft and enjoy an uninterrupted view of our world from above. Feverish with excitement, Commander Leonov instinctively cried out the first words that entered his head: 'The Earth *is* round.'

Nowadays, we are familiar with the view of our planet from space — to many it may even seem commonplace. But look again at the oceans, mountains and continents, for they are just the thinnest crust on top of a massive ball of rock and iron. How could this gigantic globe have formed from something as intangible as a cloud? The answer is a process as awesome and beautiful as the planets it left behind.

Building worlds in a computer

In 1969 a book was published in the Soviet Union by a modern-day Laplace. Called *Evolution of the Protoplanetary Cloud*, it was the work of Viktor Safronov, a mathematician virtually unknown in the West. Safronov was trying to understand how the planets formed and theorized that in the inner Solar System the dust stuck together particle by particle to form first grains, then pebbles, then boulders, then islands of rock and metal, and finally planets. He concluded that the Earth, Mars, Venus and Mercury all formed from the accretion of these smaller bodies. Furthermore, Safronov's time-scale for the construction of the planets fitted in well with the known age of the Earth. Measurements of the dwindling radioactivity of ancient rocks suggested that the Earth is approximately 4.5 billion years old. We know from the age of the oldest meteorites (rocky fossils of the original compacting grains of dust) that the Solar System is slightly older, approximately 4.6 billion years. By Safronov's calculations the inner planets would have taken 100 million years to reach full size. The sums worked perfectly, but was the theory credible?

By the early 1970s Safronov's work had reached the USA, where it hit the spot with a geologist called George Wetherill. He ran numerical simulations of Safronov's accreting Solar System on his computer. With processing power unavailable to the Soviet scientist, Wetherill was able to play with Safronov's figures, changing parameters to see if there were limits to the ways that planets formed. Time and time again the finished planets fell out at the end of his programs at about the same time — around 100 million years. The die was cast and the accretion of planets became accepted as the standard model for their creation. (The accretion theory, while at the time of writing is still widely accepted as the most likely model for formation, is not by any means the only model. Chapter 8 looks at how the discovery of alien worlds around distant stars has caused us to contemplate other theories for planetary formation.) But Wetherill's computer simulations did more than confirm the longhand algebra of Viktor Safronov — they revealed that the process of building planets is far from orderly.

Scars of chaos

On 29 March 1974 the American space probe *Mariner 10* homed in on Mercury. As soon as it was within range, its cameras kicked into life and the first ever pictures of this tiny planet took some of the mission scientists by surprise. Mercury was smothered in craters. During the

ABOVE *The grains of creation can be seen in this cross-section of a meteorite. Once the Solar System was populated with countless numbers of these lumps of rock and iron — the very foundation of the planets.*

OPPOSITE TOP *German rocket scientist Wernher von Braun (1912–77). Von Braun's V2 missiles may have failed to win the war for Adolf Hitler, but their descendants helped launch the space race on both sides of the Iron Curtain.*

OPPOSITE MIDDLE *The first V2 landed on 5 Stanley Road in Chiswick, West London. On its journey from Europe it had briefly left the atmosphere.*

OPPOSITE BOTTOM *The one planet no astronomer had been able to see before the space age: the Earth.*

encounter, the guest-room at JPL was filled with high-ranking military personnel, who commented that the planet looked like it had suffered a prolonged attack by *B-52* bombers. In a sense the generals were right – Mercury had been bombarded – but the violence that it had endured was far beyond anything that they could have conceived.

This was not the first battered world space scientists had seen. In 1964 the American *Ranger* probes (see Chapter 2) sent back the first close-up views of our Moon, which showed a similar record of violent impacts. At that time scientists had imagined that for some reason, possibly connected with the presence of the Moon, the orbit around the Earth had been unusually strewn with debris early in the Solar System's history. Then, a year later, the first pictures of Mars (see Chapter 3) showed another landscape dominated by craters. Now the pockmarked face of Mercury confirmed that there wasn't a planet in the inner Solar System that had escaped a torrent of rock raining from the sky. A rush of excitement ran through George Wetherill when he saw the first images come back – this bombardment was just what his computers had predicted.

Cosmic pinball

Wetherill's computer models had revealed that Safronov's accretionary process was, in fact, split into two distinct phases. The first he described as a period of 'runaway growth'. From a foggy cloud of dust and gas swirling around in the gravitational pull of the young Sun, Wetherill's computers showed that within just 50,000 years – an incredibly short space of time in cosmic terms – the dust in the inner Solar System would have congealed into approximately 30 planets the size of Mercury or Mars. This was a period of calm growth, where, at one stage, billions of rocks called 'planetisimals', varying in size from a kilometre to the length of Britain, were floating through the endless fog, bumping alongside each other and gently docking under the force of

Asteroids

Orbiting the Sun in a broad band stretching between Mars and Jupiter is a collection of wandering rocks called the asteroid belt. The largest asteroid, Ceres, is the size of a small moon and measures 770 kilometres across. Most are much smaller, and astronomers estimate that there are millions of odd-shaped rocks at least a kilometre across patrolling the belt. That might sound crowded, but given the massive expanse of space they occupy – a band 550 million kilometres wide – asteroids are, in reality, few and far between. In 1991 and 1993, while NASA's *Galileo* probe was on its way to study Jupiter and its moons, it caught glimpses of two of the larger main-belt asteroids. The first was named Gaspra, an odd-shaped lump 18 kilometres across. Two years later it imaged a 56 kilometre long boulder called Ida, and found it to have its own tiny moon, Dactyl. These were the first asteroids to be photographed from close range.

Not all asteroids orbit in the main belt – some swoop into the inner Solar System, occasionally passing close to the Earth. In 1989 scientists at the Arecibo Radar Observatory in Puerto Rico imaged one such asteroid, called Castalia. At a distance of more than 5 million kilometres, there is little detail to be seen on its surface, but the images clearly show two distinct lumps that appear to be joined at the hip. It is the first known example of a 'contact binary' – two rocks stuck together by the force of gravity in just the same way that the billions of planetisimals came together to form the planets.

1

In 1997, the *NEAR* (*Near-Earth Asteroid Rendezvous*) spacecraft swung past asteroid 253 Mathilde in mankind's third close encounter with these rocky remnants from the birth of the Solar System. But *NEAR*'s waltz with Mathilde was to bring a surprise. Spacecraft engineers measured the asteroid's mass through the deflection of their communications beam and found that Mathilde was unexpectedly light. After enduring millions of pounding impacts, Mathilde is not only covered with craters, but the rock within it has been reduced to a honeycomb. Mathilde is the most cratered object ever encountered – a testament to the intense bombardment that was once commonplace throughout the Solar System (see pages 28 and 29).

1 *Asteroid 951 Gaspra became the first asteroid to be photographed from space when NASA's* Galileo *spacecraft captured this image in 1991.*

2 *A three-dimensional computer model of the near-Earth asteroid 4769 Castalia. Discovered in 1989 by Earth-bound radar, it is believed that the two halves (each less than 1 kilometre across) were once separate bodies which gently kissed and never parted.*

2

Asteroids

gravity. As more of these islands grouped together, the rocks at the centre of the accreting jigsaw were squashed so hard by the mounting gravitational forces that they melted. In a remarkably short space of time these rock-piles became tiny worlds in their own right and the heat at the centre of them grew so intense that the heavier molten elements, such as iron, started separating from the rocky part and fell towards the core. If this was all there was to making planets, their faces would have ended up as smooth as billiard balls. But another stage was about to begin – one that would leave the planets scarred for life.

At some point the gravitational presence of each growing world became powerful enough to start to disrupt the hitherto neat circular orbits of its rivals. Very soon these proto-planets began to pull each other into a chaotic cat's-cradle of eccentric overlapping paths around the Sun. The sedate progress of the initial accretionary phase was quickly undone. This second stage Wetherill called 'the period of heavy bombardment'. Asteroids were raining from the sky. Giant world-smashing collisions were inevitable, and before long the orderly growth of the inner planets had turned into a fight for survival – survival of the biggest. The period of heavy bombardment lasted about 100 million years. During this time many worlds would have sported rings, early versions of the rings around Saturn, either captured piecemeal from the surrounding debris or thrown up from their surface after a gigantic impact. The energy of impacting asteroids would have left large areas of the planets' surfaces molten.

In the middle of this mayhem, one proto-planet had managed to collect more debris than the rest. One day it would be the Earth, an oasis for life, but for those first 100 million years, as bodies large and small careered into each other, our planet just had to survive.

BELOW *Meteor Crater, Arizona, is believed to be about 50,000 years old – barely yesterday on the geological clock. Despite being 200 metres deep and nearly a kilometre wide, in comparison with the country-sized impact craters found on the Moon and Mercury, this is just a baby.*

Slowly, the dwindling debris that wasn't getting eaten up by the growing planets was being thrown by their gravity into wildly eccentric orbits. Eventually, the majority of these wayward rocks strayed too close to the Sun and were dragged into its boiling core. At the end of this cataclysmic era, just four rocky planets were left in the inner Solar System: Mercury, Venus, the Earth and Mars.

One billion years later, the planets' orbits were all but cleared of the remaining debris. However, the planets are still reeling from their creation. The Earth and Mars are both tilted, spinning off vertical as a result of the countless collisions with other planetisimals that knocked them first this way then that. *Mariner 10* discovered that Mercury has a metal core far larger in proportion to its overall size than any other planet. That leads George Wetherill to suspect that Mercury would be twice its present size but for a catastrophic collision that sent the majority of its outer rocky shell spraying off into space. Among the craters found littering the surface of Mercury was one so gigantic that shock waves from it formed a mountain range clear on the other side of the planet. And of all the collisions that afflicted the rocky planets, perhaps the most spectacular was the monumental impact of two worlds that resulted in the formation of the Earth and the Moon themselves (see Chapter 2).

Despite the relative calm of the Solar System today, the impacts haven't entirely finished. There are still plenty of fragments – collectively known as the asteroid belt – left over from the formation of the inner planets and caught in a gravitational tug of war between Mars and Jupiter. It is thought that the asteroid belt is a planet that never managed to form because the fragments were always being disturbed by Jupiter's powerful gravity. If they were ever to overcome those disruptive forces, the asteroids could collect together and make a planet about the size of our Moon. Occasionally, tiny fragments broken off these asteroids, called meteorites, breach the atmosphere of our own planet and plunge to Earth. When we cut open these lumps of rock and iron we can see the initial stages of accretion laid out in their patterns of tightly packed grains. Here are the original pebbles of creation clinging together to form the building blocks of planets.

Violence beyond the snowline

Close to the Sun the early nebula was so hot that most matter could exist only as a gas. Only materials with very high melting points, such as metal and rock, could condense out of the nebula and clump together to form solid planetisimals. But at about the distance from the Sun where Jupiter now orbits, the temperature dropped low enough for water vapour and gases such as carbon dioxide to freeze into ices. Some scientists call this point the 'snowline'. Further out still, possibly near the orbits of Uranus and Neptune, it would have been cold enough for gases such as methane and ammonia to freeze. Beyond the snowline the planets formed not just from grains of rock and metal but also from snows of water and other ices, drifting amid a swirling blizzard.

In the 1970s and 1980s NASA's robotic spacecraft *Voyager 2* sailed into the outer Solar System and paid visits to the entire family of giant planets. Evidence of a period of

The orbital dance

What causes planets to orbit the Sun and moons to orbit their planets? Space is a virtual vacuum with almost nothing to slow down a speeding object. A body drifting through space would sail in a straight line for all eternity if it were free to do so. But as the planets fight to fly off in their perfectly straight lines, their course is being tugged into a circular path by the pull of the Sun, an immense entity 500 times more massive than the rest of the Solar System put together and endowed with an enormous gravitational pull. This finely balanced struggle between the planets' own momentum and the Sun's gravity was set in motion by the original spinning cloud of gas and dust around the young Sun from which the planets were created. The same is true of the motion of moons around their parent planet.

The 17th-century English physicist Isaac Newton was the first person to discover the connection between gravity and orbits, demonstrating that the same force that makes an object fall to Earth keeps the planets in motion around the Sun. He also calculated that the force of gravity lessens the further you are from an object. This discovery revealed why the planets nearer to the Sun travel round it faster than the planets that exist in the solar backwaters. Flying on the fringes of the Sun, Mercury feels the full force of our star's gravitational pull — this winged messenger hurtles through space at over 200,000 kilometres per hour, taking just under 88 days to orbit the Sun. Way out at Neptune, the Sun's gravitational pull is so weak that the blue giant ambles at a leisurely pace — Neptune travels at a tenth of the pace of Mercury and takes 164.8 Earth years to complete one full orbit of the Sun.

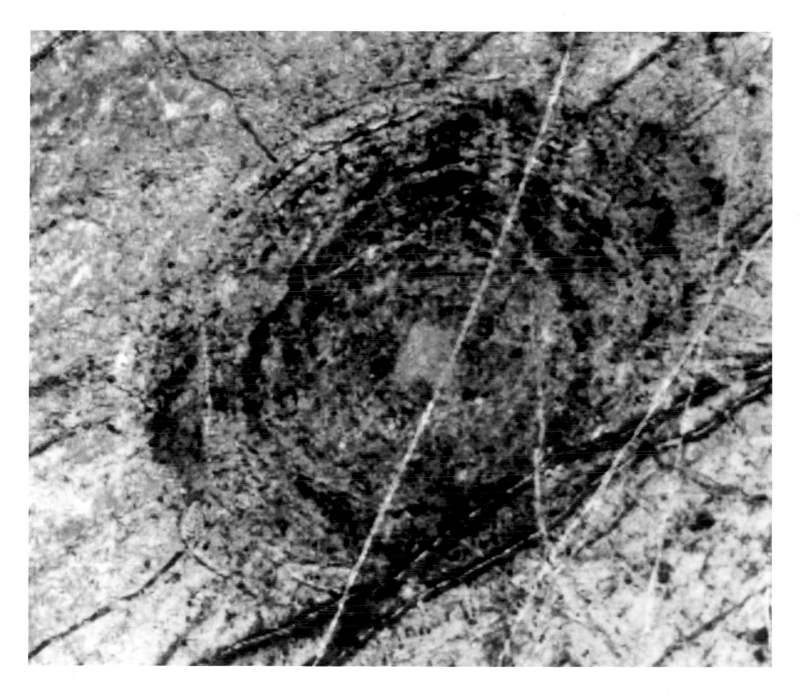

bombardment every bit as violent as the battle for survival in the inner Solar System was everywhere to be seen. The giants, being worlds of gas, would show no lasting signs of bombardment on their surfaces, but their fantastic collections of icy moons told a tale every bit as dramatic. Most of these little worlds were spattered from pole to pole with craters. The probe also sent back fabulously detailed images of Saturn's rings – possibly the relic of an icy moon shattered by a collision. Beyond Saturn, *Voyager 2* witnessed at close hand the bizarre rotation of Uranus. At some point in its distant past, Uranus was probably knocked flat on to its

ABOVE *This image of Europa, the icy moon of Jupiter, taken on 4 April 1997, shows a crater with rings stretching for 140 kilometres. The damage was probably caused by a comet or asteroid the size of a mountain.*

back by a world the size of the Earth. Neptune, too, has an exaggerated tilt of 29 degrees from the upright.

On 25 March 1993 three comet-hunters were working the late shift at the Palomar Observatory in San Diego, California. Eugene and Caroline Shoemaker and David Levy had each made many historic discoveries in their time, but the comet they discovered that night was to change their lives forever. Caroline Shoemaker was the first to recognize something amiss when she announced that an object captured on their photographic plates looked like a 'squashed comet'. In fact, they had discovered a comet broken into 21 pieces, each hurtling through space, one behind the other, in a straight line. The announcement of its discovery made page 23 of *The Times* newspaper in London. Some two months passed before the Central Bureau for Astronomical Telegrams announced that the comet, now known as Shoemaker-Levy 9 (SL-9), was on a direct collision course with Jupiter. 'This was no longer page 23 of *The Times* – this was now page one,' recalls David Levy.

By the time the comet reached Jupiter, every available telescope on Earth and in space was trained on the gargantuan planet to witness the greatest event in the history of astronomy. When it arrived, SL-9 didn't disappoint. One by one the cometary fragments, up to 4 kilometres in diameter, ploughed into the clouds at 216,000 kilometres per hour. At the impact sites mushroom clouds of scorching gas billowed 3,000 kilometres up into space, leaving huge bruises in the planet's upper atmosphere that remained visible for several months. 'This was how the Solar System was built,' enthuses David Levy, 'comets hitting planets. Jupiter grew a little that day. It has more water, more carbon monoxide, more carbon sulphide than it had before.' Comet SL-9 showed us that the final stages of accretion haven't finished. 'During the summer of 1994 there was a big yellow police fence saying, "Keep out – Solar System still under construction!"'

Impossible giants

Two decades earlier, however, when Safronov and Wetherill first ran their models for the accretion of the giant planets, they hit a snag. The further the planets were from the Sun, the longer the computer said it took to build them. Jupiter, even with all that extra snow around, needed about 100 million years. More distant Saturn, where the building blocks were spread more thinly, took even longer. That was a problem because it was common knowledge among planetary scientists that Jupiter and Saturn had to have formed much faster than these two theoreticians were predicting. The history of the Sun itself imposed a severe time limit.

Somewhere between 2 and 10 million years after the solar nebula began to collapse, the cloud of gas and dust would have undergone a radical change. The contracting mass at the heart of our nebula would have been squeezed and heated by ever increasing gravitational forces. At some point, the centre of the cloud reached sufficient pressures and temperatures to kick-start a nuclear reaction. As the nuclear generator burst into life, the Sun as we know it today turned on. A ferocious wind of charged particles blew outwards in all directions. Quite suddenly, while the countless planetisimals were busy slugging it out, the fog

cleared. This wind blew the light gases, such as hydrogen and helium, from the developing Solar System and pushed them deep into space.

Jupiter and Saturn are made almost entirely of hydrogen and helium. That means they must have formed a core of rock and ice big enough for gravity to attract and hold on to their vast quantities of gas before the Sun blew the supply away – in other words, within 10 million years. How these cores, about ten times bigger than the Earth is today, accreted so much faster than the rocky planets is still a mystery. But Jupiter and Saturn are not the only theoretical problems. Uranus and Neptune were clearly made in a different way from Jupiter and Saturn. The absence of anything like the quantities of hydrogen and helium found in Jupiter and Saturn in the two outer giants suggests that they took much longer to form and didn't finish until well after the young Sun blew the gases out past their orbits. According to the standard accretion theory, these planets would have taken billions of years to form. In fact, according to the models, poor old Neptune should still not be built today. Clearly, there has to be something wrong with the calculations.

Missing matter

The recent discovery of Jupiter-sized worlds around nearby stars (see Chapter 8) gives astronomers hope that they will soon find a giant planet orbiting a very young star – one under 10 million years old. If that happens, at least we will know that the quick-fire construction of giant planets is possible. Then the question will just be 'how?'

Perhaps it is true, as some accretion theorists suggest, that the presence of snow accelerates the gluing together of planetisimals. But the most likely possibility is that Jupiter and the other giant planets had much more material to feed on than the models of Safronov and Wetherill assume. The trouble is, it's very difficult to find out for sure just how much there was. The vast stretch of space extending nearly 4 billion kilometres from Jupiter out to Neptune is cold and empty these days. It wasn't always that way. Just as in the inner Solar System, the rapidly growing gravitational influence of the young Jupiter and Saturn would have caused mayhem amid the neatly orbiting lumps of icy debris that accompanied them. But being so much more massive than their comparatively tiny rocky cousins, the gravitational effect these hefty planets exerted on the surrounding lumps of ice was very powerful. Many of them were tossed clear out of the Solar System, the rest were flung out to orbits way beyond Pluto. There were perhaps as many as 1,000 Earth-sized worlds fighting a losing battle for gravitational supremacy with Jupiter and Saturn. Further out yet, the tilts of Uranus and Neptune suggest that they too were vying with many icy companions. Pluto was one of those companion worlds lucky enough not to be swallowed up or thrown out very far. But the rest of the matter of the early Solar System has been systematically cleared from the neighbourhood of the planets and now lies deep out in space, beyond the orbits of Neptune and Pluto. Finding that missing material – and finding out just how much of it there is – may hold the key to the mystery of the giant planets.

Comets

Every few years comets pass close enough to the Earth to be seen with the naked eye. Like frozen fireworks, they glow against the night sky, periodic visitors from the dark recesses of our Solar System. Comets have been seen for thousands of years, but what they were and where they came from remained a mystery until recent times. In the 1950s a Dutch astronomer, Jan Oort, worked out that the comets that visit the inner Solar System are just the merest fraction of a gigantic sprawling reservoir of icy fragments which stretches outwards to a distance 1000 times greater than the orbit of Neptune. They are so distant and so small as to be invisible to our most powerful telescopes, and it is estimated that there are thousands of billions of these comets, most of which were thrown out of the forming Solar System by Jupiter and Saturn.

In 1986 one of the most ambitious space adventures ever undertaken came to a climax. The Russian spacecraft *Vegas 1* and *2*, having already dropped off atmospheric probes at Venus, hurtled on to help guide a European spacecraft, *Giotto*, to within 605 kilometres of Halley's comet. Passing close to the Earth once every 76 years, Halley has long fascinated astronomers. *Giotto's* pictures were in a different league to those taken by Earth-based telescopes. They confirmed that at the heart of comet Halley lies a pitch-black nucleus of icy dust — a 16-kilometre frozen fossil from the original solar nebula. As Halley and its kin approach the Sun, they start to evaporate, sending jets of gas spurting from their surface. These tails of gas and dust can extend for millions of kilometres.

In 2004 an American probe, *Stardust*, is set to fly through the tail of comet Wild 2, sample the gas and collect the particles of dust trailing in its wake before returning them back to Earth for analysis. For the first time, scientists will have within their grasp the ancient raw material from which our Solar System formed.

1 *The nucleus of comet Halley, photographed by the European Space Agency probe* Giotto *in March 1986.*
2 *The arrival in 1810 of comet Halley was greeted with great excitement by astronomers, who had expected its return, but it still struck fear into the hearts of the populace.*
3 *Halley seen from Earth has a bright tail of evaporating ice and dust stretching over millions of kilometres.*

Back to the beginning

It took Clyde Tombaugh more than a year to discover the ninth planet. On 28 April 1998 astronomers Dave Jewitt and Jane Luu peered through the Canada-France-Hawaii Telescope that sits on top of Mauna Kea, 4,200 metres above the Pacific Ocean, and picked off five tiny new worlds beyond Pluto. It was a good night for the astronomers, but by no means exceptional. The handful of wandering stars that they had just discovered are but a fraction of the swarm of icy bodies recently observed beyond Neptune.

Ever since 1930, astronomers had wondered whether there might be more matter out beyond the ninth planet; it just seemed so unlikely that the stuff of planets should reach an abrupt end at Neptune. We now know that the Edgeworth-Kuiper Belt, named after the astronomers who first proposed its existence, is indeed the home of the icy material expelled by the giant planets. Some of these floating icebergs have orbits that periodically bring them close enough to the heat of the Sun to be melted and so become comets. Once the Solar System was perpetually strafed by these frigid lumps; now only a few remain to make occasional forays into the inner Solar System.

Scientists now believe that there are billions of these small, trans-Neptunian bodies stretching out for billions of kilometres. Pluto, far from being the last planet in the Solar System, stands at the gateway to a new realm. Dave Jewitt and Jane Luu are convinced that there are many more Pluto-sized worlds to be discovered. These worlds represent fragments of planets that never were, the very building blocks of worlds frozen in eternal purgatory in the blackness beyond Neptune. If we are to solve the outstanding mysteries of the formation of the Solar System, we shall have to go to Pluto and beyond – to the realm of the ice dwarfs.

Clyde Tombaugh died in late 1996, a pillar of the astronomical community. Throughout the era of robotic space exploration, he had been guest of honour at mission control for all the major encounters with the planets. The man who discovered Pluto was lucky enough to witness the robotic emissaries painting faces on to the wandering stars of his youth. Yet after four more decades of space exploration, his planet and its icy relatives still remain an almost total mystery.

There is a mission scheduled to send a probe to reach Pluto by around 2012. Several years ago, when the mission *Pluto Express* was first mooted, the scientists wrote to Clyde Tombaugh to ask permission to visit his planet. Tombaugh couldn't have been more proud. If everything goes to plan, the craft will subject Pluto and its moon, Charon, to a barrage of tests and send back critical information about their structure and atmosphere. It will then pick an object from the Edgeworth-Kuiper Belt and head towards the new frontier on the ultimate

voyage of discovery. Rich Terrile, the project scientist for *Pluto Express*, readily admits that he doesn't know what they will find when they reach the object in the belt. Perhaps there will be a surprise waiting in the composition of its ices, or a clue to its history in the pattern of its craters.

Some 80 years after the discovery of Pluto we might finally get to visit the world that so confounded our picture of the Solar System. Perhaps this tiny oddball and its kin may

yet yield the clues that will help us understand how Jupiter, Saturn, Uranus and Neptune came to be. Whatever *Pluto Express* might discover in the realm of the ice dwarfs, it is bound to open up fresh debates and consign accepted theories to the rubbish heap. If 40 years of space exploration have taught us anything, it is that nothing can prepare us for what we will find.

moon

IN AN INSTANT THE FLICKERING digital displays scattered around mission control freeze and the assembled mass of *Apollo* flight technicians in Houston takes a deep breath. Some 380,000 kilometres away, three men travelling at 8,000 kilometres per hour have just disappeared behind the Moon. If anyone in Houston wants to send a message to the astronauts of *Apollo 8* in the next 45 minutes, they are out of luck. The trio plummet into the blackness of the far side of the

Moon. So far on their journey they've
been deprived of a glimpse of their
target, but now, as they slip into
darkness and their eyes adjust to the
gloom, the sky fills up with millions of
stars. From the window of the
capsule, Bill Anders sees an arc of
total blackness ahead of them. Then,
just three minutes before the service
module rocket is due to light up in
front of them and slow them into an
orbit just 110 kilometres above the
Moon, the black arc edges into
pristine sunlight. 'Oh, my God,'
Anders murmurs to his crewmates.
'Look at that!' The three peer down
through the thick glass to see
mountains jutting up from the
long shadows of the lunar morning.
They are seeing something humans
have dreamed of for centuries.
And for another half-hour on this
Christmas Eve of 1968, no one
else will know.

three hundred and fifty-nine years before Bill Anders, Frank Borman and Jim Lovell became the first men to look down on the Moon from close quarters, Galileo Galilei turned his newly built telescope up to the bright disc in the sky above Florence. With the miracle of magnification, he saw the edge of the Moon was not divinely smooth and round, but bumpy and imperfect. That night the father of modern astronomy had his own private moment in which he alone knew that the Moon was a real world, made of rock, perhaps not so different from our own.

Not long after that, the skies revealed more of their mysteries. Casting around for more wonders to capture with his new toy, Galileo focused his lens on giant Jupiter, and nestling next

PAGES 38–9 *Best seat in the house: Apollo 8 astronauts snapped this picture of the Earth rising above the surface of the Moon.*

to it he saw four little points of light. As he tracked their movements from night to night, he was astonished to see that all four were circling the distant planet. Our Moon, it appeared, was not alone. The Solar System had a whole family of second-class citizens – objects whose orbits went around planets, not the Sun. There was not one Moon, but many moons.

Since 1610, the number of satellites known to grace the Solar System has grown from five to more than 50, many of them discovered between 1979 and 1989 by the *Voyager* space probes which journeyed to the outer planets – Jupiter, Saturn, Uranus and Neptune. These moons are some of the most fantastic worlds we have ever seen. Icy globes spattered with millions of craters, hunks of rock that seem to have been blasted apart before falling back together, a moon with geysers of frozen nitrogen, another with an atmosphere thicker than the

ABOVE LEFT *Galileo Galilei (1564–1642) made these sketches of the Moon nearly four centuries ago.*

ABOVE RIGHT *Galileo presents his telescope to the Muses and points out Jupiter and its satellites, along with Venus and Saturn.*

Earth's. But the very existence of our own Moon is perhaps the most surprising of all.

The Earth is the only rocky planet to have a substantial moon. Mars has two tiny satellites, Phobos and Deimos, only 17 and 9 kilometres long respectively. They aren't even round; they are most likely asteroids that wandered too close to the path of the red planet and became captured in its gravitational draw. Venus and Mercury have no moons at all. Why the Earth should have a moon a full quarter of its width is one of the deepest mysteries of our Solar System. To ancient stargazers the Moon was a celestial mirror: its surface appeared mottled because it was reflecting the continents and oceans of the Earth. Galileo scoffed at that idea. When he looked at the dark markings on the Moon's bright face, he thought they were real seas and called them *maria*. As telescopes improved, astronomers discovered that the *maria* were riddled with small craters – they were, in fact, drier than any desert on Earth. But scientific advances have never quelled the Moon's mystical draw.

The Moon has always been a lure. But in the age of space travel it became a just-possibly-attainable goal. The race the USA and the Soviet Union fought to get to the Moon is a story that has already passed into legend. On a summer night at the close of the 1960s, two Americans walked on Sea of Tranquillity. But behind that unforgettable picture of an astronaut planting the stars and stripes on a barren, dusty plain lies a tale of scientific adventure of unrivalled proportions. What kind of world is our nearest neighbour? How old is it and where did it come from? The date 20 July 1969 was an important day in history – it marked the beginning of our journey towards a true understanding of the Moon.

C h o o s e   t h e   M o o n !

We choose to go to the Moon! We choose to go to the Moon in this decade and do the other things – not *because they are easy, but because they are hard.*

US President John F. Kennedy, from a speech made at Rice University, Houston, Texas, 12 September 1962

Ever since the Soviets filled the void of outer space with the blast of *Sputnik*'s radio static in 1957, the heavens had become the symbolic battleground of the superpowers. While most people fretted over the threat of assured mutual annihilation, lunar scientists couldn't have asked for a luckier break. The USA's most popular president had made a commitment to put a man on the Moon before the 1960s were out (to match a perceived Soviet effort to do the same), which meant they would soon get to see another world at close hand. But the Americans had a lot of catching up to do.

Three years to the day before Kennedy made his stirring speech to a gathering of Texan students, a craft emblazoned with a hammer and sickle blasted off on the first successful journey to the Moon. On 15 September 1959 the Soviet Union's *Luna 2* probe crashed into the Sea of Serenity. Although little scientific knowledge stemmed from it, the wrecked lump of metal in a pile of lunar dust scored the first point in the superpowers' race to the Moon. To drive it

OPPOSITE *Shades of grey: our Moon's bright surface is littered with strange dark patches called* maria *or 'seas'. One of the goals of a manned Moon mission was to find out how the Moon and its* maria *formed.*

The **binary** planet

To look up at the night sky and behold the pale crescent of the Moon it is hard to believe that this serene object is in reality a 70-billion-billion-tonne ball of rock careering around the Earth at 3,700 kilometres per hour. No other planet in the Solar System, besides little Pluto, can boast a satellite of such relative importance. The Moon and Earth are close enough in size and space that some astronomers prefer to think of them together as a single 'binary planet', for each globe is inextricably linked to the other.

Study the Moon's face for a month and you will notice its apparently changing shape. The waxing and waning is no change in the Moon itself, merely a trick of the light. As the Moon makes its 27.3-day orbit around the Earth, it is lit at different angles by the Sun.

suck up a bulge in the ocean about 1 metre high. On the other side of the planet from this point there is a similar bulge: this part of the ocean hangs slack, feeling the weakest force from the Moon. The Earth spins once every 24 hours, but the bulges in the ocean stay put, locked to the Moon, so each point on the ocean shore passes through two high tides and two low tides each day.

It was George Darwin, son of the famous biologist, who first suggested that the Earth and Moon system may not always have been the way it is now. According to his Victorian theories, the sloshing tides were gradually halting the Earth's spin, and the Moon was drifting away from the Earth. Darwin studied tidal markings on ancient corals and found that when they

During Full Moon, the Sun is behind the Earth, at Half Moon the Sun is directly to the side of our satellite as we look at it. The faint face of the New Moon gets no direct sunlight, only the pale glow of Earthshine – the Sun's rays reflected off our planet.

Look more closely and you will also see that the Moon always presents the same face to the Earth. One side of the Moon happens to be slightly heavier than the other, and the Earth, 80 times as massive as the Moon, uses its powerful gravity to keep that heavy side facing it all the time.

But the Moon does not sit in quiet obedience. It pulls back on the Earth and creates our daily ocean tides. The point on the Earth directly under the Moon feels the strongest force, which is enough to

had been alive – several hundred million years ago – the Earth's day was only 20 hours long. Another British scientist, Harold Jeffreys, trawled through records of the times of lunar and solar eclipses back to Babylonian days and found that the Moon did indeed seem to be getting further away – he judged the rate to be 3 centimetres a year.

When *Apollo 11* landed on the Moon in 1969 it confirmed Jeffreys' theory and measured the rate of drift to be 3.8 centimetres a year. The effect of the slow Moon drift is to lengthen the day by two-thousandths of a second each century. This might not sound like much, but it adds up; when our binary planet first formed, the Moon was ten times closer to the Earth than it is now, and an Earth day lasted a mere six hours.

firmly home, one week after the crash-landing, Nikita Khrushchev presented a replica of *Luna 2*'s pennant to President Eisenhower during the first visit of a Soviet premier to the USA. If this put a dent in American pride, worse was yet to come.

At a New Year party in Moscow that same year, a group of young space scientists was popping champagne corks. To be drinking fine French wine in the depths of a Russian winter seemed even more remarkable to them than their miraculous achievement: the champagne was their reward for bringing the world the first ever pictures of the far side of the Moon.

Several years earlier a rich French eccentric had publicly declared that he would give a case of his finest champagne to whoever could show him the side of the Moon that always points away from the Earth. When *Luna 3* did just that, in October 1959, the Frenchman made good his promise. Although most of the bottles were scooped up by greedy *apparatchiks* at the Soviet Embassy in Paris, Boris Chertok, one of the engineers on the *Luna 3* team, fondly remembers that by the end of the year the alcoholic bubbles were popping on his tongue. 'It's not everyone who can say they've drunk French champagne in return for the Moon!' he recalls.

After two successful Soviet probes, NASA's efforts seemed paltry. *Pioneer 4* was its only successful lunar mission, but it had carried no camera and had flown by the Moon at a distance of around 60,000 kilometres. The USA was way behind in the space race, and this time it couldn't even claim the Soviet mission had been scientifically worthless: there was something important on the far side of the Moon. Piecing together the 17 fuzzy images from *Luna 3*'s camera made a picture that was not much more than a blob. But the image was almost completely free of the dark *maria* – Galileo's 'seas' that dotted the near side. The far side of the Moon was unendingly grey and chock-full of craters. This other world, hanging tantalizingly close to the Earth, was clearly a place worthy of exploration.

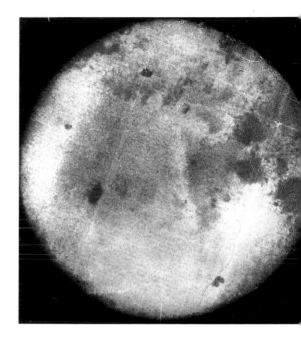

ABOVE *The first picture of the far side of the Moon, pieced together from the 17 photographs taken by* Luna 3 *in 1959. On the right you can see part of the near side.*

If not for politics...

American geologist Jim Head remembers seeing an advertisement in a technical journal in the early 1960s showing a full-page picture of the Moon. Below it a caption read: 'Our job is to think our way to the Moon and back.' When he finally made it on to the growing team of geologists that was being recruited for NASA's new *Apollo* programme, Head knew he'd be making history. The astronauts being considered for the first Moon landings were military test pilots chosen for their exceptional flying skills. Head would help decide where the astronauts would land and train them to know what to look for on the lunar surface. They would become his eyes and hands on the Moon. It would be the highlight of his scientific career. While the astronauts were destined to bask in public glory, the geologists would have the chance to see how their first new world measured up to the Earth. They had scores of unanswered questions.

If the mystery of the far side of the Moon was one more item on the geologists' long-standing list of unsolved riddles, the origin of the *maria* themselves was up near the top. By the end of the 19th century, most astronomers and geologists agreed that the lunar seas were, in all probability, vast flows of lava that had seeped up through cracks in the Moon's crust. But where were these fissures? Earth-based telescopes could only see objects on the Moon larger than a few hundred metres across – too coarse to pick out the sources of these floods of molten rock. Sending spacecraft to the Moon would afford an up-close and personal view.

Another debate that had been raging since Victorian times was the origin of the innumerable craters that make the Moon's face look like a pock-marked mess. Some astronomers were apt to think of the craters as the tops of volcanic cones. But others, particularly those who had visited places like Meteor Crater in Arizona, suggested they were the scars of asteroids striking the Moon at tens of thousands of kilometres per hour.

A thousand years earlier there was no doubt which camp a group of British monks would have sided with. In 1178 Gervase of Canterbury wrote of how he and his brothers were gazing up at the Moon when they saw something very strange. All of a sudden, a small distinct crescent appeared just over the rim of the Moon. Gervase saw that crescent 'split in two…a flaming torch sprung up, spewing out…fire, hot coals and sparks. Meanwhile, the body of the Moon which was below writhed…and throbbed like a wounded snake.' The monk had no explanation for what he saw, but it now seems likely that he had witnessed the impact of a meteorite which threw a large cloud of debris above the surface and caught the light of the Sun.

As telescope images of the Moon improved, the idea that the craters were the result of impacts began to gain hold. The shallowness of the bowls, the bright rays fanning out incredibly far from the biggest craters, the sheer number of them…everything pointed towards the lunar surface being blasted from above rather than being pushed up from below. If that was the case, the Moon was an ancient world whose surface hadn't changed much for billions of years, perhaps barely at all since the day it was formed. The Moon might be a relic – the key to unlocking the history of the early Solar System, history that has been wiped clean on Earth by wind, water, volcanoes and, of course, life.

Closer

SEQUENCE OPPOSITE *Photographs taken by the kamikaze* Ranger 9 *probe, as it made its final fatal approach to the Crater Alphonsus. Cameras strapped to the* Rangers *snapped away furiously before the probes ploughed into the surface. No matter how close it got,* Ranger 9 *saw only crater upon crater.*

OPPOSITE LEFT *Map of the Moon, showing the area of* Ranger 7's *approach in 1964. The last few of the probe's pictures showed 1,000 times more detail than the best telescopic views from Earth.*

If a manned landing were to be successful, it was clear that many more unmanned probes would first have to perform a thorough reconnaissance of the lunar surface. As the 1960s ticked away, the Russians and the Americans raced to design, test and fly a whole flotilla of new Moonships.

NASA chose the simplest approach to get closer to the Moon, with a series of kamikaze probes called *Ranger*. They had been designed at the Jet Propulsion Laboratory (JPL) in Pasadena, California. JPL would later develop the phenomenally successful *Mariner*, *Viking* and *Voyager* probes, but the newly founded centre for the unmanned exploration of space got off to a terrible start. As the *Ranger* probes hurtled into the lunar surface at several thousand kilometres per hour, they were to radio back pictures of the Moon from TV cameras strapped to their noses. The first five probes, however, were utter failures; only *Ranger 4* actually hit the Moon. President Kennedy's 1969 deadline (honoured by his successors) left no room for such mistakes, and the mission manager was hastily fired. The new chief simplified the design of *Ranger 6*, unbolting all non-essential scientific experiments. It was now nothing more than a guided missile with a camera on the front. *Ranger 6* worked just fine, impacting at precisely the intended spot on the Moon; unfortunately the camera shorted out and didn't take a single picture. Finally, on 31 July 1964, in the seconds before *Ranger 7* slammed into the Sea of Clouds, mankind saw the first-ever detailed pictures of another world.

The last few *Ranger 7* pictures showed 1,000 times more detail than the best telescopic views from Earth. As the craft fell closer and closer to the surface, more and more smaller and smaller craters appeared. However closely you looked at it, the Moon was thoroughly peppered with the scars of meteorites. It was a sight that enchanted both scientists and the world alike. Growing in confidence with every mission, the JPL chiefs eventually allowed the nation to witness their attempts as they happened. On 24 March 1965 *Ranger 9* produced the first live TV show from the Moon as all the major US networks covered the last few moments of its headlong plunge into the crater Alphonsus.

On the other side of the Iron Curtain, there were no plans for a TV spectacular any time soon. In parallel with JPL's early difficulties, the Soviet space effort had suffered a string of utter failures. In the period between *Luna 3*'s triumphant journey and *Ranger 9*'s live pictures, only

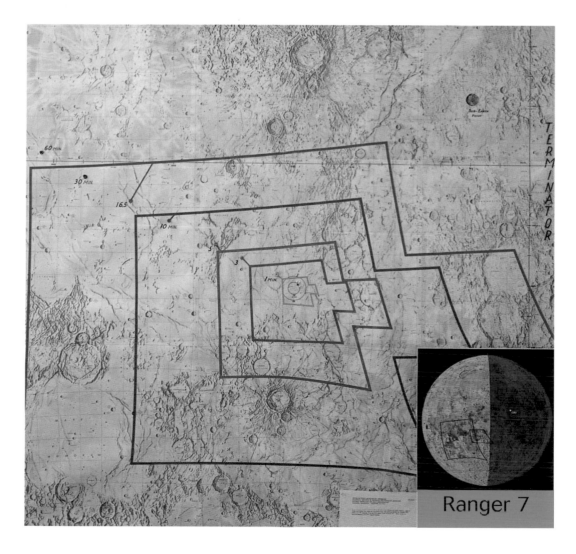

Ranger 7

BELOW One of the early Ranger *probes built to crash-land on the* Moon. *After* Rangers 1—5 *failed to hit their target, the design was simplified. The rather intimidating saw-toothed ball at the top was, unfortunately, one of the sacrifices.*

three more of their lunar probes had made it off the ground, and none of them returned any useful information. But now the Russians were striving to perfect the technically complex task of making a controlled landing on the surface of the Moon. This meant that the probes had to be fitted with retro-rockets to slow them down enough to settle gently on the lunar surface. They made their first attempt in May 1965, but *Luna 5*'s retro-rockets malfunctioned and the probe smashed itself to pieces in the Sea of Clouds, joining the wreckage of its earlier rival, *Ranger 7*. *Luna 6* missed the Moon altogether. Then, as 1965 drew to a close, *Lunas 7* and *8* began a new pile of scrap metal in the Ocean of Storms. The surface of the Moon was an elusive target.

Race to the Moon

1

USSR

1959 September 15 *Luna 2* crashes into the Moon and becomes the first spacecraft to reach another celestial object.

1959 October 7 *Luna 3* sends back the first-ever pictures of the far side of the Moon.

1961 April 12 Yuri Gagarin becomes the world's first space traveller, orbiting the Earth once, spending 1 hour and 48 minutes in space.

1966 February 3 *Luna 9* sends back the first panoramas from the lunar surface.

1966 April 4 The Moon gains its first artificial satellite as *Luna 10*, designed to take measurements of lunar gravity, radiation and magnetism, drops into its orbit.

1968 September 21 *Zond 5* is the first craft to go to the Moon and come back in one piece.

1969 July 3 The giant *N-1* booster rocket, needed to put cosmonauts on the Moon, fails for the third time, effectively putting an end to the Soviet manned mission to the Moon.

1969 July 21 An attempt to bring a lunar rock sample back to Earth with a robotic craft is scuppered when *Luna 15* crashes into the Sea of Crises.

1970 September 24 The first Soviet piece of the Moon arrives back on Earth, courtesy of *Luna 16*.

1970 November 17 An unmanned Moon rover called *Lunokhod 1*, controlled from the Crimea by a team of operators, takes soil samples and analyses them on the spot.

1976 August 22 The final Soviet Moon mission of the 20th century, *Luna 24*, brings back another haul of Moondust, which it digs up from 2 metres below the surface.

USA

1962 February 20 John Glenn becomes the first American to spend a significant amount of time in space – during his 4 hours and 55 minutes aboard *Friendship 7* he made three orbits of the Earth.

1964 July 31 *Ranger 7* crash-lands on the Moon with its cameras clicking away. These first close-up pictures of the Moon are a thousand times better than previous telescopic views.

1966 August 14 *Lunar Orbiter 1* goes into orbit around the Moon. It sends back the first picture of the Earth as seen from the Moon.

1967 April 19 *Surveyor 3* lands on the Ocean of Storms. Its robotic arm is the first to dig into the lunar surface and sample the soil.

1967 November 9 The giant *Saturn V* rocket, capable of lifting the 52 tonnes of equipment needed for a manned mission to the Moon, makes its maiden flight without a hitch.

1968 December 25 The crew of *Apollo 8* makes the first manned flight around the Moon. They make ten orbits at a height of just 110 kilometres.

1969 July 20 *Apollo 11*'s Neil Armstrong and Buzz Aldrin become the first men on the Moon.

1970 April 13 After an explosion on board, the crew of *Apollo 13* is forced to abort the third lunar landing mission. All three astronauts make it back safely to Earth.

1971 July 30 Dave Scott and Jim Irwin take the first drive on the Moon in *Apollo 15*'s lunar rover.

1972 December 7 *Apollo 17* carries the last two Americans to walk on the Moon. One of the moonwalkers is Jack Schmitt, the first geologist to visit another world.

1990 December 9 The *Galileo* spacecraft, en route to Jupiter, takes the first good images of the south pole of the Moon.

1994 February 21 *Clementine* enters lunar orbit. It makes an intense study of the lunar poles, possibly detecting signs of ice in craters at the south pole.

1998 January 7 *Lunar Prospector* begins a three-year-long orbital reconnaissance mission to the Moon.

1 NASA's 'original seven' astronauts. Al Shepard (second from right) was the first to go up in a Mercury capsule, spending just over 15 minutes in space. John Glenn, the first American to orbit the Earth, is third from left.

2 Yuri Gagarin in training for the first human trip into the void of space. In 1961 Vostok 1 took him around the world in 89 minutes.

The race quickens

The *Ranger* pictures confirmed what scientists had suspected for more than a decade – the Moon was covered in a layer of dust. Aeons of pelting by meteorites – most only the size of grains of sand – had pulverized the rock on the surface of the Moon into a layer of fine particles. Nobody knew how thick this dust might be. Thomas Gold, a professor at Cornell University in upstate New York, warned that the grey powder could be tens or hundreds of metres deep over the entire surface of the Moon, and that any astronauts trying to land on it would likely sink without trace.

The risk of ignominiously losing the first men on the Moon in a bizarre death by 'drowning' was one that neither NASA nor the Soviets were about to take. Before people landed, robotic probes would have to test the footing. The soft landing was a vital skill that both superpowers needed to master quickly. In February 1966 the Soviets finally got it right. *Luna 9*'s retro-rockets fired 46 seconds before landing and the craft made a perfect touchdown in the Ocean of Storms. Ground controllers had to wait four minutes before they knew that the on-board equipment had survived unscathed. But right on cue, *Luna 9*'s high-pitched radio squawk filled the room, and the first broadcast from the surface of another world had begun. *Luna 9* returned three historic panoramas of the Moon's surface. From a vantage point safely on top of the compact dust, it saw a slightly drab world strewn with angular rocks, some as small as a few millimetres across, others as big as small houses.

If *Luna 9* put the Soviets back on track, two months later they pushed firmly ahead of the Americans when *Luna 10* became the first craft to orbit the Moon. *Luna 10*'s official mission was purely scientific: to measure the amount of radioactivity emanating from the surface of the moon and determine what it was made of. The readings were very similar to volcanic rock on the Earth. But *Luna 10* also cleared two more obstacles to a successful manned mission. The very fact that the craft achieved a steady circular path around the Moon proved that the Soviets were capable of doing the same with a manned craft – a crucial step in any landing attempt. And its Geiger counter proved that the amount of radiation in lunar orbit was not too much for a cosmonaut to withstand.

Lunas 9 and *10* sent the Americans into a panic. The press was spreading the word that the West was slipping behind. NASA reacted swiftly with a headlong rush to the Moon. Between June 1966 and January 1968, five American *Surveyor* craft successfully soft-landed on the Moon. They were far more sophisticated than *Luna 9*: they took pictures of the surface, dug the soil with automatic scoops, and saw the Earth shining brightly in a black sky above an airless, grey horizon. One craft, *Surveyor 6*, even made double sure about the safety of lunar dust by restarting its rocket engines after touchdown to

rise against the Moon's feeble gravity and hop 2.5 metres to one side. *Surveyor 6's* camera then looked back at the original spot and noticed no deep robotic footprints in the soil.

Hand in hand with the *Surveyors*, five *Lunar Orbiter* craft swooped into low orbits over the Moon. By January 1968 they had taken high-resolution photographs of most of the lunar surface. Back in Houston, everything was falling into place for the *Apollo* manned missions to the Moon. Landing sites had been selected. Wernher von Braun's behemoth of a rocket, the 110-metre-tall *Saturn V,* had passed its first test flight without a hitch. And the lunar module, the part of the *Apollo* spacecraft that would carry men on the final leg of the journey down to the surface from lunar orbit, had already flown in space.

ABOVE *Muscovites flock to the news-stands to read all about* Luna 9's *triumphant soft-landing on the Moon in 1966.* Pravda's *front page (right) may have been scooped, but its pictures were sharper and didn't suffer from being accidentally vertically stretched like the intercepted images that the* Daily Express *had published (see opposite).*

T i m e   f o r   t h e   m e n

As 1968 drew to a close, the Americans suddenly began to feel less comfortable about the lead they had opened up. Intelligence reports from inside the Soviet Union hinted that a manned *Soyuz* space capsule, the one everyone expected to carry cosmonauts to the Moon, was ready to fly. Then, in September, came the news that an unmanned craft, *Zond 5*, had made a figure-of-eight trip from Earth around the Moon and back, carrying turtles and even transmitting pre-recorded messages from cosmonauts to test communications. That really unsettled the *Apollo* managers. Until then they had been happy to hear that the Soviet answer to the *Saturn V* – the *N-1* booster – had been plagued with troubles. Lacking that kind of

RIGHT *The 42-metre-tall first stage of the* Saturn V *rocket in the Vehicle Assembly Building at the Kennedy Space Center in Florida.*

BELOW *Next to* Surveyor 3*'s footpad are signs of a bumpy landing. This picture, taken during the* Apollo 12 *moonwalk, shows just how thick Moondust is.*

firepower, a Soviet manned mission to land on or even orbit the Moon was out of the question. But launching a manned *Soyuz* craft on a 'free-return trajectory', using the Moon's gravity to bring the spacecraft back to Earth using a simple figure-of-eight flight plan, was easily within reach of the present Soviet booster. And *Zond 5* had proved they knew how to pull it off.

NASA hadn't planned to send men to orbit the Moon until the fourth manned *Apollo* mission. Now they made a snap, and some said reckless, decision: directly after *Apollo 7* had made the first manned flight in Earth orbit, men would fly to the Moon, fire their retro-rockets, settle into orbit around our neighbouring world for a day and then return to Earth. The crew of *Apollo 8* would have the honour of being the first men to blast themselves completely free of the Earth.

On Christmas Eve 1968 Frank Borman, Jim Lovell and Bill Anders floated just 110 kilometres above the Moon. So far from home, Lovell described to mission control mankind's first new world in a masterpiece of understatement: 'The Moon is essentially grey…no colour. Looks like plaster of Paris.' It seemed an odd place to want to visit so badly. But the world's press made more fuss over *Apollo 8* than any other space mission. Man had broken the symbolic bond with the Earth. Walking on the Moon was not far away now.

When *Apollo 8* returned to Earth, everything seemed to change. Suddenly, Kennedy's goal to land on the Moon before 1969 was out seemed achievable. And although the Soviet booster was powerful enough to perform the figure of eight, it stood no chance of matching *Apollo 8*'s feat of ten lunar orbits, let alone going one better and landing on the Moon. In their congratulatory telegram to the Americans, a whiff of Russian defeat was already in the air when they claimed that the Moon race had never even happened except in the minds of the Americans. In May 1969 *Apollo 10* went back to the Moon again, this time taking a lunar landing module down to just 15 kilometres above the surface. Then, on 3 July, the Soviet effort effectively died when a test firing of the *N-1* booster ended for the third time in a row in a giant explosion on the launch pad.

BELOW *Cramped quarters. The crew of* Apollo 8, *from left to right: Bill Anders, Jim Lovell and Frank Borman, in position for a Moonflight.*

Thirteen days later, with little over five months to go before Kennedy's deadline, *Apollo 11* set off on the first mission to land men on the Moon. Neil Armstrong, Edwin (Buzz) Aldrin and Michael Collins carried with them not only the hopes of a patriotic nation, but also those of the dozens of scientists sitting in the geology back room at mission control in Houston. Robotic probes had scanned and sampled the lunar rock and soil, but the geologists would never really understand the Moon until they could hold its rocks in their hands.

Two ships in the night

While *Apollo 11* was making its four-day journey to a date with destiny, a flurry of panic went through mission control in Houston. A small Soviet craft was already in orbit about the Moon: what was it doing there? Were the Soviets trying to steal their thunder? Some of the scientists with energetic imaginations thought it might even be a sabotage mission. Under a shroud of secrecy, the Soviets had sent another unmanned probe, but this time there was a difference. *Luna 15* was making a two-way trip. If all went well, it would return to Earth with a precious cargo – about 100 grams of lunar soil. What's more, a robotic sample-return mission had a striking advantage over a manned one: the Soviet lander could choose its landing site by scientific merit, rather than opting for the safest, flattest terrain. Although the laws of physics meant that *Luna 15* would arrive back a day later than Armstrong and Aldrin, the 'Moonscooper' would ensure the Soviets came in a thoroughly respectable second in the long race. And they could always argue that they were obtaining lunar soil without risking the lives of cosmonauts. *Luna 15* was due to land in the Sea of Crises, just west of *Apollo 11*'s destination, the Sea of Tranquillity.

Meanwhile, Soviet geologist Sasha Basilevsky spent his July waiting by the shores of the Caspian Sea, ready for *Luna 15*'s splashdown back on Earth. He had been primarily responsible for choosing the Soviet lunar landing sites and now he was desperate to get his hands on that small drill core from another world. The Americans, as things turned out, had no need to worry. Basilevsky remembers acutely his disappointment when he heard that *Luna 15* had gone out of control at the end of a four-minute descent into the Sea of Crises. At the time *Luna 15* crashed, *Apollo 11* had been safely on the surface for several hours. 'I was pleased for the Americans,' Sasha remembers. 'How could a scientist not be glad a man had landed on the Moon? But I was envious, too.' The first Moon rocks on Earth would be studied in Houston, not Moscow.

But the journey of *Apollo 11*'s lunar module, the *Eagle*, had been anything but uneventful. As Armstrong and Aldrin stood side by side in the cabin, riding down on top of the bright flame of the *Eagle*'s descent engine, their computer began to experience problems. A series of blaring alarms assaulted the ears of the two men, and when they were just 300 metres above the surface of the Moon, Armstrong realized that the computer was guiding them to a landing site that was strewn with giant boulders.

Mustering all his test-pilot experience, Armstrong snatched control from the automatic programme and pitched the lander forward to stop the rocket from braking its forward movement. He gave the booster a few extra spurts to slow their fall and try to get the *Eagle*

OPPOSITE *Not long for this world: the crew of* Apollo 11 *begins its historic mission from Cape Canaveral, Florida, on 16 July 1969.*

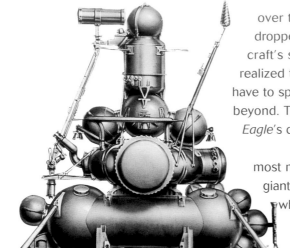

over the boulder field to a place that looked smoother. The fuel gauge dropped to 8 per cent, and from the corner of his eye Aldrin could see the craft's shadow racing up to meet them. At the last moment, Armstrong realized that his chosen spot was still too rugged and that the *Eagle* would have to spend its last dregs of fuel guiding them over a crater to a clear plain beyond. Then, with less than 30 seconds of fuel left, the contact light in the *Eagle*'s cabin came on. They were down!

A few hours later Armstrong uttered the words that must be the most memorable spoken this century: 'That's one small step for man, one giant leap for mankind.' In fact, Armstrong fluffed the line he meant to say, which was to have been: 'That's one small step for *a* man, one giant leap for mankind.' No matter, he had better things to think about now. He was walking on the Moon.

A field trip on the Moon

The surface is fine and powdery. I can kick it up loosely with my toe.
It does adhere in fine layers like powdered charcoal to the
sole and sides of my boots.

Neil Armstrong's second sentence from the
surface of the Moon, 20 July 1969

For politicians, Armstrong's first sentence, spoken with feet firmly on airless dust, signalled the end of the USA's journey to the Moon. His second sentence was for the geologists and it began their voyage of understanding. When Armstrong climbed out of the hatch of the *Eagle*, the grounding in geology that he had received back on Earth had already equipped him with the equivalent of a master's degree. Fifteen minutes later, Aldrin would poetically describe the same scene with the words 'magnificent desolation', but for the geologists in the back room at Houston, Armstrong's technical description was more like what they wanted. For the next two and a half hours the eyes of two astronauts on the Moon would be theirs too, and there was so much they wanted to see.

Through the electronic fuzz of a primitive TV camera, the world watched two men struggling to get used to the peculiar ease of walking in one-sixth of Earth's gravity, trying to plant an American flag in Moon dust without it toppling over, and posing for the most remarkable tourist snaps ever taken. But the mission priority was to get a sample of the Moon to bring back to Earth, and Armstrong had already bundled a small sack of rocks into his pocket before Aldrin had even stepped out into daylight. This 'contingency sample' meant the crew wouldn't go back empty-handed if there were a sudden emergency during man's first Moonwalk. Now, as he loped across this barren landscape dotted with small craters, Armstrong was captivated by the showers of dust arcing up in a perfect fountain from his every footstep, and by a footprint as sharp as if it had been made in talcum powder. Aldrin couldn't believe how the minerals glinted so brightly when raw sunlight bounced off the rocks. The rocks were dark, looking very much like volcanic lava on the Earth — basalt. They were full of holes, which Armstrong took to be signs of bombardment by tiny meteorites. He was wrong about that: the holes were later shown to be where gas had oozed out of the lava as it was solidifying, just as in pumice stone.

By the end of those 151 minutes on the Sea of Tranquillity, the first two explorers had set the stage for a revolution in our understanding of the Moon. They set up a scientific station to listen for tremors beneath the powdery volcanic surface. They laid out an array of mirrors off which scientists would bounce a laser beam and measure more accurately than ever before the distance from the Earth to the Moon. Most important of all, they were heading home with 21 kilograms of rocks and dust, specimens that would be the object of intense study by no fewer than 130 eager geologists.

OPPOSITE TOP *Artist's impression of the Soviet* Luna 16 *spacecraft, which landed on the Moon, in the Sea of Fertility, on 20 September 1970. It brought samples of lunar rocks back to Earth. Its predecessor,* Luna 15, *crashed on the Moon*

OPPOSITE *Edwin (Buzz) Aldrin quipped about making sure not to lock the hatch behind him as he climbed out of the* Eagle. *Neil Armstrong was still the only human ever to walk on the Moon when he took these photos of Buzz.*

RIGHT *A boot crunches into the lunar dust. Only the gentle erosion of micrometeorites will affect this mark. It should last a million years or more.*

OPPOSITE *Seize the moment: Buzz Aldrin poses next to the American flag. When he and Armstrong blasted off the Moon just two hours later, the flag fell over.*

How to get to the Moon and back

No single effort of humankind, aside from the mass D-Day landings of the Second World War, can claim to have taxed people's ingenuity, effort and pay-cheques quite as much as the task of sending two men to walk on the Moon. At its peak, NASA's *Apollo* programme employed 400,000 people and took about a dollar a week from every single taxpayer in the USA. It was a journey of unprecedented complexity and precision. The target was a ball in space

380,000 kilometres away and 3,500 kilometres across. Just hitting the Moon would have been a challenge, but nothing as simple as a straight shot was possible. No rocket was powerful enough to blast three men and all their necessary equipment in one direct shot from the Earth to the Moon. The trip had to be broken down into stages. For each *Apollo* mission to be a success well over a dozen separate rocket firings had to happen right on cue. Here's how it happened:

Mission time

Lift-off: Hold tight The five first-stage engines have been firing for three seconds before the giant restraining clamps release the 110-metre-tall *Saturn V* rocket. Riding on 160 million horsepower, within 40 seconds it has broken the sound barrier. After two and a half minutes, the first stage has drunk all its half million gallons of kerosene and raised the trio of astronauts 64 kilometres high. G-forces peak at four and a half times normal.

11.5 minutes: Floating After a six-minute burn, the second stage falls back to Earth, and the third stage spends just under three minutes getting the spacecraft into orbit 184 kilometres above the Earth.

2 hours 45 minutes: No turning back The third stage fires again,

making the Translunar Injection. After just five minutes and 18 seconds the astronauts are travelling at 38,800 kilometres per hour and are headed to the Moon.

3 hours 24 minutes: Shunting carriages The astronauts in their Command/Service Module (CSM) cut themselves free, pull away a few dozen metres and turn the CSM around to face the spent third-stage rocket. Nestled inside the end of the rocket is the Lunar Module (LM). The CSM pilot guides his ship back in to dock with the LM and pulls it free of the rocket.

5 hours: Slow cooking As the CSM and the LM speed to the Moon in the unceasing glare of the Sun, precautions are taken to avoid one side of the spacecraft overheating while the other side freezes. The ship goes into 'barbecue mode', making one spin on its axis every hour to ensure even cooking.

3 days 4 hours: In orbit again After the long cruise through outer space, a few minutes' blast from the CSM's engine is enough to slow the spacecraft into a parking orbit 110 kilometres above the Moon.

4 days 4 hours: Three into two The CSM pilot is left all alone as the commander and Lunar Module pilot separate their craft and fire the LM's rocket for two minutes to bring them down to 15 kilometres above the barren, cratered landscape below.

4 days 7 hours: All the way down With the LM flying on its side at 6,000 kilometres per hour, the descent-stage rocket fires up again and the LM begins a long, slow brake down to the surface. At 2,300 metres high, the LM tilts upright and the astronauts ride down to the Moon on top of a bright flame, controlling their rate of

Lunar lift-off + 11 hours: Going home The Service Module comes to life again, and three minutes later has broken free of the Moon's grip and is heading back to Earth.

Lunar lift-off + 3 days: Skip, skip, burn Just short of the Earth's atmosphere and travelling at a lightning speed of 38,800 kilometres per hour, the Command Module jettisons the Service Module and the astronauts brace themselves for the most jarring phase of the journey. The Command Module would be fried to a crisp if it tried to plunge right through the atmosphere, so instead it bounces twice off the atmosphere, each skip slowing it down until it is safe to descend. Even so, the crew still experiences a G-force of six, barely able to breathe, let alone move. Three kilometres up from Earth a giant red-and-white parachute unfurls and drops them gently into the Pacific Ocean.

descent. Four minutes later they should be down, but if anything goes wrong, there's all of 60 seconds' extra fuel built in.

Lunar lift-off: Getting legless After spending anything from 21 hours to more than three days on the Moon, the two Moonwalkers begin the first leg of their journey back. The now-useless legs and descent rocket of the LM are left on the Moon as the top half blasts back up for a rendezvous with the CSM 110 kilometres above the surface.

Lunar lift-off + 3 hours: Reunion The CSM pilot shakes hands with his two crewmates, the LM's ascent stage is sealed off and at the flick of a switch is cast off into a slowly decaying orbit, destined eventually to crash-land on the Moon.

1 *Armstrong leads his crew off on the first stage of the journey – the 13-kilometre bus ride to the Saturn V rocket.*

2 *Point of no return: Apollo 15 blasts off from pad 39A at Cape Canaveral.*

3 *Testing: Apollo 9 astronaut Dave Scott gets out of the Command/Service Module, which is docked to the Lunar Module in Earth orbit.*

4 *Good communication was paramount in all Apollo missions. Here the antenna on the lunar rover of Apollo 17 is pointed back at Earth.*

5 *Splashdown on the Pacific. One of the three parachutes on Apollo 15 failed to open, but the crew survived.*

6 *All over bar the shouting. Armstrong, Aldrin and Collins get a heroes' welcome on New York's Broadway.*

How to get to the Moon and back

No stone left unturned

When the astronauts got back to Earth, they were immediately quarantined by NASA, just in case they had brought back any strange diseases from the Moon. The rocks were quarantined too, and subjected to a barrage of cutting, drilling, slicing, vaporizing and X-raying that Armstrong and Aldrin were grateful to avoid. The first thing everyone wanted to discover was the age of the rocks. The Sea of Tranquillity was one of the vast, dark basins on the Moon that had at some point in the past been flooded with lava. Dating the rocks would tell when that had happened. From dating meteorites, geologists already knew that the Solar System was 4.6 billion years old. Assuming that the Moon formed at around the same time, the volcanic plains were expected to be older than 3 billion years, since the Moon was a small body and would have cooled off fast, according to conventional geologic wisdom. But a few scientists were hoping that the samples might be much younger, maybe only 1 or 2 billion years, and that the Moon had been volcanically active until much more recently.

ABOVE *Neil Armstrong, Buzz Aldrin and Mike Collins smile for their wives while trapped for 21 days in a quarantine trailer after their return to Earth.*

The conservative camp won the day. The samples were all about 3.7 billion years old – as old as the most ancient rocks found on Earth. The Moon was indeed the geological fossil they had expected it to be. The mineral make-up of the rocks was broadly similar to the basaltic rocks found near volcanoes on Earth – mostly iron and silicon. One aspect, however, was very strange. There were far fewer light elements in the samples – elements that boil away at lower temperatures than the heavier elements such as iron or titanium. And, strangest of all, there was absolutely no water in the Moon rocks. It was almost as if you could make a Moon rock by taking an Earth rock and roasting it until all the water and the lighter minerals had boiled away.

Their appetites were whetted, and lunar geologists desperately wanted more. Four months later, *Apollo 12* landed on another of the lunar *maria*, this time the dark spot visible on the left side of the Moon, the Ocean of Storms. Astronauts Pete Conrad and Alan Bean got to spend a little longer walking around up there – eight hours – and they brought back a haul of rocks weighing 34 kilograms. These samples would answer a new question: did all the dark patches on the Moon form at the same time? It turned out that the Ocean of Storms was half a billion years younger than the Sea of Tranquillity. If the *maria* had formed at such different times, it seemed likely to the geologists that they were the scars of giant meteorite impacts, apocalyptic events that (as described in Chapter 1) were all too commonplace in the first billion years of the Solar System's history. The impacts gouged out huge basins, which then 'bled' with lava. Traces of the violence of the impacts showed in the rings of mountains still visible around the *maria*, like ripples frozen on a pond. If it were somehow possible to have stood on Earth and patiently watched our satellite for its first billion years, we would have seen the Man in the Moon slowly taking shape. First the left eye, then the right, and finally the nose would have seeped black lava on to the Moon's bright face.

The primordial crust

In the summer of 1971 geologists were eagerly awaiting a new type of mission. NASA had come under fire as the US Congress demanded something more than nice TV pictures for the $24 billion the Moon landings were chewing off the nation's budget. A combination of indifference, ineptitude and bad luck was calling the whole programme into question. *Apollo 13* never reached the Moon after a life-threatening explosion in the service module shortly after leaving Earth orbit. *Apollo 14*'s enduring legacy was the first live image of man playing golf on the Moon. Commander Al Shepard, who had been the USA's first man in space ten years before, readily confessed to not caring much about rocks, but his mission nearly finished *Apollo*. The rocks brought back were poorly documented: the geologists were as furious as some of the politicians in Congress. Either it was time to do some real science up there, or stop going altogether. NASA's response was to rewrite the mission rules for *Apollo 15*. The

ABOVE *There was one consolation prize for the Soviets after the historical landing on the Moon of* Apollo 11. *In the winter of 1970, they sent the first roving vehicle to the Moon. There were no cosmonauts riding around on* Lunokhod 1 *(above left), but it was controlled remotely from the Crimea (above) for nearly 11 months, taking 20,000 photographs and collecting 500 soil samples.*

next flight would send astronauts to live on the Moon for three days. They would have improved backpacks for longer Moonwalks. They would even have a brand new electric-powered rover to cover more ground. Most of all, the next two Moonwalkers, Dave Scott and Jim Irwin, would see their geology field trips and study periods rise to top priority in their training schedules.

Apollo 15's target was a winding 'river' on the edge of the Sea of Rains called Hadley Rille. On our waterless Moon it seemed likely that these sinuous channels (there are about a dozen of similar size dotted around the surface) had been carved by raging torrents of lava. Hadley Rille is more than 1 kilometre wide and 400 metres deep. If the astronauts peered down inside it, they might be able to see layers of lunar bedrock, while around them they might find the volcanic debris that would surely have sprayed up when this hellish river was last active. Dating those rocks would tell geologists more about when the Moon was last active, but landing at Hadley did have one more advantage. Rising up from that corner of the Sea of Rains were the 5-kilometre-high lunar Apennines. Directly above the rille towered the 3,500-metre summit of Hadley Delta. If Scott and Irwin could drive the rover just a little way up the slope, they might be able to snatch a truly ancient rock from the lunar highlands and that might help tell the story of the Moon's formation.

The Genesis Rock

Back in the 1960s, ideas about how planets formed were still in their infancy. As explained in Chapter 1, the theory of accretion was only just starting to be worked out at the time. Many geologists still believed that planets slowly gathered themselves together as cold lumps of rock — volcanoes and lava only came later when heat from natural radioactive decay had built up inside for hundreds of millions of years. But a new idea gaining ground was that the planets were born hot from the incessant pounding of one giant meteorite strike after another. In fact, these early worlds would have been so hot that they would be entirely covered by an ocean of lava. Such a dramatic theory needed evidence, and the Moon was the perfect test-bed. Unlike the Earth, our diminutive Moon had never been very active, so pieces of the original crust, almost as old as the Solar System itself, might well survive up in the highland areas.

Scott and Irwin had been told to be on the look-out for a piece of whitish rock called anorthosite. According to the geologists' calculations, if the Moon had once been coated with a global lava ocean, anorthosite, a light mineral, would have floated to the top and cooled first. It would have formed the original bright crust of the Moon, which was only later stained with the dark *maria* lava flows.

On 1 August 1971 two space-suited field geologists drove away from Hadley Base aboard *Rover 1*. Driving along the edge of the ancient river of lava, they headed up into the foothills of Hadley Delta. Looking back at the valley behind them, they saw boulders the size of houses. So did Houston, which was being fed live colour pictures from the Moon, a far cry from the fuzzy black-and-white images of Aldrin and Armstrong two years before. Ahead, the Apennines were almost golden in the sunlight of the lunar morning. It was the astronauts' second day in the field, and before it was out they would have made two incredible discoveries.

They were about 100 metres high and nearly 5 kilometres away from the lunar module, now little more than a small metal speck. Heading back downslope, they stopped at the rim of a crater called Spur. Suddenly, Irwin saw something strange – a rock that sparkled green light back towards him. He lifted his gold visor to check the colour was right – it was indeed a lump of green, glassy rock, full of bubbles. Scott bagged the sample, wondering how this rock could have got its peculiar hue, but his mind was soon on other things, for ahead of him lay a small white rock. Could this be what they were looking for? Scott turned the sample over carefully in his hand. It was covered with dust, but his gloved fingers were wiping it clean. And then he saw them – large white crystals just like the ones he'd been trained to recognize on Earth. There was no doubt in his mind: this was a piece of anorthosite. Scott's voice crackled excitedly on the radio: 'I think we found what we came for.'

They had yet to peer down into Hadley Rille's snaking depths, but the discovery of anorthosite was the high point of the mission. Back on Earth the rocks came good. The green one was a glob of tiny glass bubbles, probably frozen spray from the raging torrents of lava that must once have been common around Hadley Base. The sample was 3.7 billion years old. But the white lump clocked in at 4.5 billion years old – surely as old as the Moon itself – and it was indeed a piece of anorthosite. It was clear proof that the Moon was once swimming in an ocean of

lava, and this little rock immediately shot to stardom as the most important piece of the Moon on Earth. The geologists called it 'The Genesis Rock'.

A g e o l o g i s t o n t h e M o o n

On 7 December 1972 the last *Saturn V* booster cleared the tower at Kennedy Space Center in Florida. Strapped in the left seat next to two Navy fighter jocks was an unlikely astronaut. Harrison (Jack) Schmitt was the first geologist to head, hammer in hand, for a field trip on the Moon. Congress had curtailed NASA's funding, and although missions had been planned all the way to *Apollo 20*, this flight – number 17 – would be the end of the line. With Schmitt's trained eye, the final mission had the best chance of rounding off the *Apollo* programme on a scientific high note.

By the time Schmitt and his commander, Gene Cernan, stepped out of the lunar module in the valley of Taurus-Littrow, scientific understanding of the Moon had already become pretty extensive. Samples returned by *Apollo* astronauts had allowed geologists to test the ages and compositions of hundreds of lunar rocks. They were all discovered to be at least 3 billion years old. It was now clear that the *maria* had flooded with lava after the early days of giant meteorite impacts, and the highlands were full of old rock largely unchanged since the Moon's early history. But there was still one nagging question: had the Moon been active at all in the last 3 billion years, or was this truly a cold, dead world?

Taurus-Littrow, a young-looking volcanic plain between two ancient mountain ranges, might have the answer. Scattered around it are several dark craters looking suspiciously like recently active volcanic vents. It was here that Schmitt and Cernan drove their rover in the closing hours of the *Apollo* programme, hunting for signs of fresh ash or lava. Just half an hour from the end of their second Moonwalk, they parked their vehicle at the rim of Shorty Crater. From the air, this light-coloured bowl could be seen to have a dark rim around it, perhaps the result of a volcanic explosion, but there was no way to tell from the air – the halo could well have been thrown out by a meteorite impact. Schmitt walked over to a large boulder near the rim and was preparing to take a panoramic photo when something caught his eye. Amid all the dark grey dust around him he saw a bright streak. 'There is *orange* soil,' he said in an incredulous tone. It was so bright that Cernan could see it from several metres away. To Jack Schmitt's eyes, this looked just like a volcanic deposit from Shorty, and a recent one too. If it had been sitting outside the crater for more than a few million years, the Sun's radiation would have slowly darkened it and the impacts of tiny meteorites would have covered it over.

In the geology back room, people were jumping for joy at the thought of having found a recent volcanic vent. For the Moon to have been active in the last few million years was hard to believe, but it was incredible news. Running low on time, Schmitt and Cernan raced to gather a few samples of the soil they were sure would cause a storm back home and perhaps even force a complete re-write of lunar history.

ABOVE *The Genesis Rock: a 4.5 billion-year-old piece of anorthosite, the ancient crust of the Moon, was the prize find of the* Apollo 15 *mission.*

OPPOSITE TOP *Astronaut Dave Scott works up a sweat, drilling a 2-metre deep hole in the Moon in search of rock samples.*

OPPOSITE BOTTOM *Scott admires his find, the piece of anorthosite, now lightly sealed in a case of pure nitrogen.*

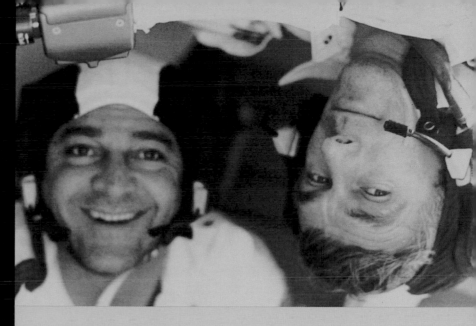

MAIN PICTURE *The* Saturn V *rocket carrying* Apollo 17 *blasts into the night sky at Cape Canaveral on 7 December 1972.*

INSET Apollo 17 *astronauts Gene Cernan and Ron Evans have a head to head during their weightless journey. This was the last manned mission to the Moon.*

Back on Earth, the orange soil was found to be a collection of volcanic glass beads that must have sprayed out of the Moon in a tall fountain of fire before cooling and settling back down into the dust. But the age was a letdown – 3.7 billion years old. The volcano had died a long time ago. The Moon had deceived even Schmitt. The soil appeared young because it had only recently been revealed by an impact. Shorty was an impact crater after all, and the Moon remained a fossil world.

If that was a disappointment to Schmitt, it wasn't all bad news for planetary scientists. Since the *Apollo* missions, the Moon has become the yardstick for working out the age of many other bodies in the Solar System. Since the Moon has been largely inactive for the last 4 billion years, it still shows nearly all the scars from the early days of the Solar System, when asteroids struck with far more regularity than they do today. When they first dated rocks

brought back from the Moon, geologists realized they had the makings of a cosmic clock. The heavily cratered parts of the Moon were such and such an age, the sparsely cratered areas were so much younger, and areas with moderate amounts of cratering were an intermediate age. Since the Moon, Mars, Mercury and Venus were all being bombarded by similar amounts of meteorites, scientists can apply this technique to places where getting rocks is out of the question for the time being. Thanks to *Apollo*, we now have a way to tell roughly how old any part of a rocky planet is, be it the dried-up valleys on Mars or the plains of Venus. Just count the number of craters on that surface and compare it with the Moon.

An odd couple

By the time Schmitt and Cernan splashed down in the Pacific, on their return from the Moon in the closing days of 1972, the *Apollo* missions had brought back a total of 382 kilograms of Moon rock. Also by then, the unmanned Soviet probes *Lunas 16* and *20* had made the demise of their predecessor, *Luna 15*, a distant memory. Between them they had brought back 130 grams of lunar soil, and in 1976 *Luna 24* would scoop up another 170 grams. Mankind now knew much more about what his nearest neighbour was made of, but there was still one enduring mystery. How did this strange pair of worlds, the Earth and the Moon, come to be?

In the early 1970s a young scientist called Bill Hartmann was pondering the origin of the Moon. At the time, there were only two theories thought to be at all possible: either the Earth had captured the Moon, or else the Moon had detached from the side of the Earth when our planet was a ball of rapidly spinning lava. Hartmann wasn't overly impressed with either of these theories. Neither seemed to make sense in light of all the new data on lunar rocks that had just been brought back.

Hartmann was puzzled by the contradictory nature of the Moon rocks. In chemical make-up they were remarkably similar to Earth rocks, and certain rare radioactive elements showed up in exactly the same proportions in both. That seemed too good to be a coincidence and certainly seemed to tip the scales in favour of those who believed the Moon had been spat out by the Earth. But, as Hartmann saw it, that couldn't be right either, since

LEFT TO RIGHT *Last days on the Moon: Apollo 17 commander Gene Cernan test drives the lunar rover, while Jack Schmitt tries out his new toy – a special rake for sifting pebbles from Moondust. Later, Cernan goes for a spin across the valley of Taurus-Littrow, and Jack Schmitt samples a rather large boulder.*

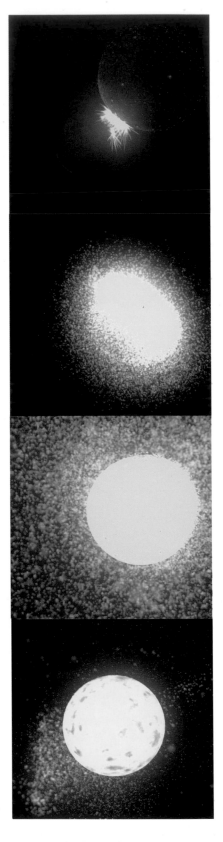

most light elements in the Earth's crust were far less common in the Moon rocks. Water, the lightest of the bunch, was completely missing from all the American and Soviet samples. So how could the Moon have formed from the Earth?

Suddenly, Hartmann had an idea. He looked at the face of the Moon and was reminded what a violent place the early Solar System was. There were traces of huge collisions everywhere he looked. Couldn't the Moon itself have been created in a giant collision, only this time a collision of some giant object with the Earth?

In 1974 an unknown, inexperienced Bill Hartmann got up to tell his theory to a group of scientists gathered at Cornell University. Feeling more than a little self-conscious, he told the story of an apocalyptic moment when a planet at least the size of Mars hit the Earth. He explained how the incredible heat generated would turn the surfaces of the colliding worlds almost instantly into molten lava. Giant clouds of super-heated lava and dust would be thrown into space – part rogue planet, part Earth. What remained of the convulsing Earth would swallow up the rest of the rogue planet. The ejected molten debris would, he calculated, coalesce incredibly quickly – in perhaps as little as 1,000 years – to form the Moon. When he had finished, Hartmann was terrified. Soon the audience would start ripping his theory apart. First to raise his hand was Alastair Cameron, one of Harvard's top planetary scientists. This was it, thought Hartmann. But as the shock of his own daring wore off and he tuned into the words issuing from Cameron's mouth, he was astonished to find that this guru was agreeing with him. Cameron, too, thought the Moon had to have been created when two worlds collided.

It is by no means proven that the Moon was formed in such a moment of biblical violence, but all the pieces fit together: the Moon has no water because it all boiled off in the great heat of the collision, and the Moon's gravity was too weak to hold on to its tenuous atmosphere. The same happened to all the lighter elements in varying degrees. On the other hand, the similarities between Earth and Moon rocks are naturally explained by their being fleetingly mixed together as two planets tore each other apart. Another measurement made on the Moon adds weight to the collision theory: *Apollo*'s seismic detectors also discovered that the Moon's metal core is tiny. If the Moon formed from the ejected crusts of two impacting planets, the heavier elements deep in their centres would in all likelihood have been swallowed up by the Earth, depriving the Moon of heavy elements from which to form its own core.

The Earth survived the collision, the other planet did not. Now it's forced to accompany the Earth for ever, a pale remnant of its former self. It tried to leave the Earth's bonds, but failed. In the moments after the collision, pieces of the Earth were thrown up into space, and some of them were drawn to the growing body of the Moon, lured towards it by its gravity. Four and a half billion years later, the Moon still has a draw. From 1959 to 1976 it worked its charm to pull from the Earth a few dozen spaceships and tempt 21 men to fly around its now lifeless corpse. The exploration of the Moon is by no means over, but it may never again be quite so much the focus of mankind's aspirations.

LEFT *A computer animation simulating the apocalyptic collision of the bodies that became the Earth and the Moon.*

The new Moon

For sheer passion, drama and adventure nothing is likely to rival the *Apollo* missions to the Moon for many decades to come. But mankind has been back to the Moon since those heady days of the late 1960s and early 1970s, and what the new unmanned probes may lack in glamour, they are making up for with some surprising discoveries.

In 1989 the *Galileo* spacecraft set off on a roundabout journey to the gas giant Jupiter. *Galileo* had cameras and detectors that put those of the *Apollo* era to shame. As it hurtled past the Moon, it looked back at the far side and saw something incredible near the south pole. It was a giant impact crater, stretching 2,500 kilometres across. A few *Apollo 15* images had given a hint that something like this might lurk to the south of its orbital reconnaissance area, but the Aitken Basin defied all expectations.

In 1994 NASA joined forces with the US military to send a small satellite, *Clementine*, to the Moon. It mapped the entire Moon in visible, ultraviolet and infrared light. But *Clementine* really made the news two years later when something strange was revealed at the south pole. It had bounced a radar signal off a crater whose floor was always in shadow. The reflected 'blip' was twice as strong as is normal for Moon rock, and seemed to be exactly what would be expected from ice.

After *Apollo* had found the Moon to be utterly dry, the idea of ice trapped in permanent night on the deep floors of polar craters was so enticing that in 1998 another NASA mission, *Lunar Prospector*, went back to try to confirm the finding. *Prospector* has no cameras — they would be of no use for looking into the black crater floors — but it can detect hydrogen embedded in the lunar surface (water is H_2O, so wherever there is water you will find hydrogen). The first results do seem to confirm lots of hydrogen at the lunar poles, although at the time of writing many scientists are sceptical that the findings really indicate water. If there is water on the Moon, it must have been scattered there by the impacts of comets and asteroids. Away from the heat of the Sun, ice could remain hidden in craters for billions of years.

1 Clementine's *image of the south pole of the Moon.*
2 *In this* Galileo *image of the Moon, the pinkish areas are highland rocks, blue and orange regions are volcanic lava flows, and the dark blue patch is the titanium-rich Sea of Tranquillity.*

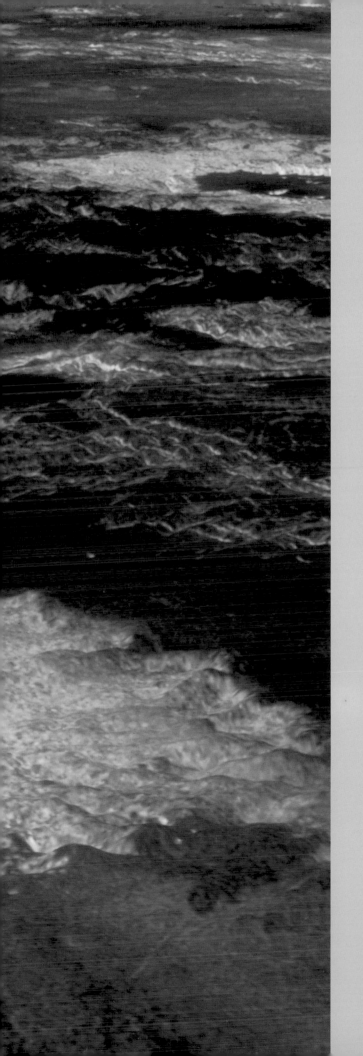

terra firma

IN THE VACUUM OF SPACE, no sound accompanies the uncoupling of the kettle-shaped probe from its mothercraft. *Venera 9* drifts down towards the cloud tops of Venus; it's now at the point of no return. And if the refrigerated mass of heavily reinforced steel — the pride of Soviet engineering — can't feel the tension, back in the Crimea a small group of scientists huddles nervously around computer screens. Success

will transform them into heroes of the Soviet Union overnight. The on-board TV cameras have been checked; everything is working. Within a couple of hours *Venera 9* should send back the first pictures from the surface of another planet. Sixty-five kilometres from touchdown the protective shield is jettisoned. A metallic parachute jolts the descent from free-fall to something survivable. Clouds of sulphuric acid start to buffet the probe and its temperature surges violently. Conditions will get worse too, but *Venera 9* is made of sturdy stuff. When it hits the surface, it might as well be landing on Hades. The crushing pressure is 90 times that on Earth. The temperature is 450 degrees Centigrade. Under such extreme conditions, the scientists know the probe won't last long — it has to work fast. Explosive bolts send the camera lens cap tumbling to the surface; a thick glass lens stares into the unknown.

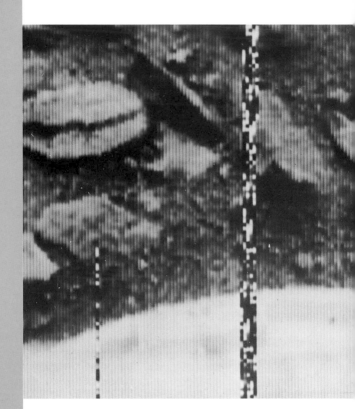

t 8.28 a.m. on 20 October 1975 scientists in
mission control in the Crimea received the signal to say that *Venera 9* was down and systems
were clinging on. But they would have to wait an agonizing 15 minutes for the image to be
sent from the surface to the orbiting mothercraft, which would then relay it to Earth. Eventually,
the paper started to crawl out of a primitive black-and-white facsimile machine. The image
was distorted by the thick lens of the probe's camera, but the surface of Venus appeared no
gloomier than an overcast day in Moscow. In the distance, the light sky and the dark grey
surface separated along a crisp horizon. Between the horizon and the craft they saw a terrain
sparsely dotted with random slabs of volcanic rock on a bed of fine soil. Mankind had just got

PAGES 74–5 *A sidelong view of
Aglaonice crater on Venus. Two more
meteorite impact craters are visible in
the background in this radar image·from
NASA's* Magellan *probe.*

its first view from the surface of another planet. But even before the paper had time to yellow
and curl, another world was to reveal itself to robot eyes.

20 July 1976: in Viking mission control at the Jet Propulsion Laboratory (JPL) in
California, a furious argument had been raging for weeks. NASA's attempt to match Soviet
success on Venus with a landing on Mars was running into trouble. *Viking 1* had been circling
the red planet for a month, looking for somewhere safe to land, and had found nowhere. A
bicentennial touchdown on the Fourth of July had been scrapped when the first images from
orbit of the planned landing site showed knobbly outcrops, craters and pits on what was
supposed to be a smooth plain. By now, mission engineers had scanned more than 800
pictures, an area larger than Texas. They finally found a flat expanse, 900 kilometres away

ABOVE LEFT *Mankind's first view of
Venus, courtesy of* Venera 9 *in 1975.*

ABOVE RIGHT *A year later, in 1976,*
Viking 1 *took this first picture of the
surface of Mars.*

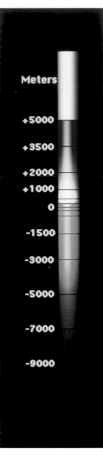

Meters

+5000

+3500

+2000
+1000

0

-1500

-3000

-5000

-7000

-9000

from the original spot. But their nerves were frayed. Would the lander hit a boulder, land on a slope, or just sink into the dust? Just after midnight on the seventh anniversary of *Apollo 11*'s lunar landing, mission manager Jim Martin bit the bullet and commanded the orbiter to jettison the lander. Within half an hour, dead or alive, it would be on the surface.

It was 4.13 on a summer afternoon on Mars when *Viking 1* came gently to rest. Another lens cap bounced off the surface of another world. Within an hour, the first black-and-white picture was coming back from Mars, strip by strip. At an agonizingly slow pace, a tiny patch of ground and the intrusive metal foot of the lander appeared. The anxiety of the preceding hours was blown away in an explosive cheer and before the scientists even had time to scrutinize the patch of alien soil, a panoramic landscape arrived. 'Look at the beautiful rocks!' screamed Martin. He saw a dry, sandy terrain strewn with angular clumps of volcanic rock, a lot like a desert here on Earth. The next day, colour images began to stream in and the world got its first view from the surface of the red planet. And, yes, it was red.

It had been an unforgettable nine months. Earth's nearest neighbours were no longer just points of light in the night sky – they were real places, landscapes you felt you could almost step into. But if snapshots of Venus and Mars had been reassuringly familiar, the geology beneath – the creator of those scenes – remained a mystery. Anyone wanting to understand how these worlds worked on the inside would have to have something to compare them with. The geology of the Earth would be the yardstick by which all the other rocky planets would be measured.

The planet that eats itself

Four and a half billion years ago the Earth was a molten ball of rock, and ever since that time it has been gradually losing heat. We can see the signs of its fiery birth even today at volcanoes, which spew liquid rock out of the ground. Its internal fire is also stoked by the natural nuclear power station inside our planet: vast quantities of uranium, thorium and radioactive potassium pump out a steady source of heat. Although the surface has now cooled enough to form a solid crust 30 kilometres thick, the insides remain molten to varying degrees. In the middle of our world is a large and dense core, part solid and part liquid metal: an alloy of iron and nickel. Outside the core, most of the rest of the Earth consists of semi-molten rock called the mantle. Neither fluid nor solid, the mantle has a consistency rather like toffee.

The slow movement of this glutinous rock under the crust is what shapes our landscape. Where it pushes up the crust, we see mountains. Where it sucks the crust down, the depression fills with water and we see oceans. All this had already been discovered by scientists many decades ago, but the Earth hid one more secret from us right up until the dawn of the space age.

It was the threat of global war and the resulting exploration of the ocean beds of the Atlantic and Pacific by nuclear-powered submarines that triggered the discovery. Hidden below kilometres of water was an immense snaking chain of volcanoes that almost encircled the globe. New areas of crust were constantly being formed on either side of this undersea ridge, and as the ocean beds grew, the continents were gradually being pushed around. Suddenly, an observation made many years before – that the coastline of South America and Africa seemed to want to fit together like pieces of a jigsaw – made perfect sense.

The Earth's crust is not a solid shell but is broken into several pieces. It soon became clear that these pieces, or 'plates', are moving ever so slowly in different directions around the Earth. Like water boiling in a pan, convection currents within the mantle rise to the surface, cool and then sink back down again. Driven by heat from the Earth's core, these gently churning semi-molten rocks ever so slowly push the plates around the planet.

Where the edges of plates rub together, we see fault lines and feel the ground tremble with earthquakes. Where plates crash headlong into each other, mountain ranges form or, in some places, one plate slips underneath the other and heads back down into the boiling mantle. Where plates are sliding apart, mostly on the ocean floor ridges, lava routinely erupts, generating new crust to fill the gap. In other places, rising plumes of hot rock in the mantle punch holes right through the middle of the plate. As the plate slowly slides across these hot spots, chains of volcanoes emerge, like those we see in Hawaii. This gentle but inexorable movement of the Earth's crust has been called 'plate tectonics' and is the trademark of our planet's geology.

BELOW *Fault zone: a river's winding course is abruptly broken by the San Andreas Fault in California, which cuts left to right across the middle of this aerial photograph.*

Were the other rocky worlds alive like the Earth? Did their ground also tremble and spit fire? Until the dawning of the space age, the Earth was all we knew. Imagine, then, with what rapt anticipation geologists waited for news from those other cosmic experiments in rocky planet formation across our Solar System.

Volcanoes

Volcanoes blow when the pressure of hot rock welling up from below gets too much for them. The vast majority of volcanoes girdle the Pacific Ocean on the so-called 'Ring of Fire'. Here, the heat comes from the friction of one tectonic plate straining to slide underneath another. Deep down, some rock melts and forms a pool of molten rock, or magma, which begins to rise like a bubble in a lava lamp.

All magma contains trapped gas, and as the magma is forced upwards, the pressure on it eases off and the gas begins to bubble out. The more gases there are in the magma, the more powerful the eruption. The volcanoes in the Ring of Fire tend to have very gassy magmas, and their ashy eruptions are called *plinian* (after the Roman writer Pliny, who saw the cloud of ash blast out of Vesuvius). When the magma has a small amount of gas in it, as in the Hawaiian volcanoes, the eruption rises only a few tens of metres above ground at most, in a fine spray of droplets of lava – a *fire fountain*. Sometimes the gas comes out in belches, creating bombs of lava – these are called *strombolian* eruptions after the Italian island volcano where the effect was first seen. When eruptions occur at underwater volcanoes, the enormous pressure of water stops the gas bubbling out at all, and the magma oozes out like toothpaste from a tube, forming *pillow lava*.

The surface of the Earth is covered with over 1,000 volcanoes which are known to have erupted at least once since humans have been around to talk about it. Some of them have put on spectacular shows of pure devastation. The blast of the 1883 eruption of Krakatoa, a volcanic island nestled between Java and Sumatra, was heard nearly 5,000 kilometres away; ash was blown even further afield and the island was all but gone when the dust finally cleared. On 18 May 1980 Mount Saint Helens on the northwest coast of the USA blew its top in a similar way, right in front of TV cameras. First it sent a searing, all-destroying wall of hot ash out from its flank, then it exploded upwards, lofting an enormous dust cloud 20 kilometres into the sky.

1 A fire fountain shooting from Mauna Loa on Hawaii.
2 Mount Saint Helens, in Washington, USA, erupting in 1980.
3 Pillow lava seeping out of an underwater volcano, deep in the Pacific Ocean.

Volcanoes

Venus, veiled sister

Throughout the 18th century astronomers observed bright and dark patches on the face of our closest neighbour and thought they were seeing Venusian landscapes. The Italian astronomer Francesco Bianchini published maps of the 'oceans' and 'continents' of Venus in 1726. In a spirit of adventure, he'd named two of them after Italy's famous explorers of new lands, Christopher Columbus and Amerigo Vespucci. To him, Venus was the next frontier. It was our sister world, almost the same size as the Earth and the nearest to us in the family of planets that orbit the Sun.

But the markings on Venus were deceptive. In fact, what the astronomers had seen was not the surface of Venus at all – it was a thick veil of cloud. This dense, reflective layer that first attracted their gaze had actually concealed the planet's surface from their view. Astronomers continued to scrutinize the planet in the hope that a break in the cloud might afford them a tantalizing glimpse of the world that lay beneath. Had they known that its clouds were nearly 30 kilometres thick, they would have saved their time.

By the late 1950s, there was at last a way of seeing through the clouds: radar. From the Goldstone tracking station in the Mojave Desert in California and the giant Arecibo Observatory in Puerto Rico, radio signals set off for Venus. The radio waves passed through the clouds as if they weren't there, then hit the surface and scattered. Some bounced back to Earth, a few even returning to the Mojave and Arecibo, and they were just enough for the scientists' purpose. Long before spacecraft were dispatched to Venus, this crude interplanetary radar built up an indistinct and fuzzy picture of our sister planet.

That picture was an aphrodisiac to those who gazed at the planet of the goddess of love. To the north was a vast, smooth plain the size of Australia and an area that looked like a mountain range bigger than the Himalayas. But Earth-bound radar was too clumsy to get detailed information from the surface of Venus. A further frustration for the early radio astronomers was a side-effect of the planet's incredibly slow (243-day) rotation: at its closest to the Earth, Venus always shows us the same face. We would never be able to see the far side from here. As soon as it was technically possible, Venus became the priority destination for a stream of Soviet and American craft.

In the 1960s the first robot emissaries to Venus radioed back some disturbing news. Conditions beneath the creamy clouds were unimaginably terrible. The American *Mariner 2* in 1962 and the early Soviet *Venera* probes recorded surface temperatures of over 400 degrees Centigrade – hot enough to melt tin or lead – and an atmosphere dense enough to crush a deep-sea submarine. As will become clear in Chapter 6, this planet's atmosphere was no sister of ours, but was the geology of Venus so alien?

LEFT *NASA's Pioneer Venus made this first crude image of the surface of Venus in 1980, piercing the thick clouds with a radar 'depth finder'.*

OPPOSITE *Another cloudy day on Venus. The details in the planet's atmosphere only become apparent when viewed in ultraviolet light.*

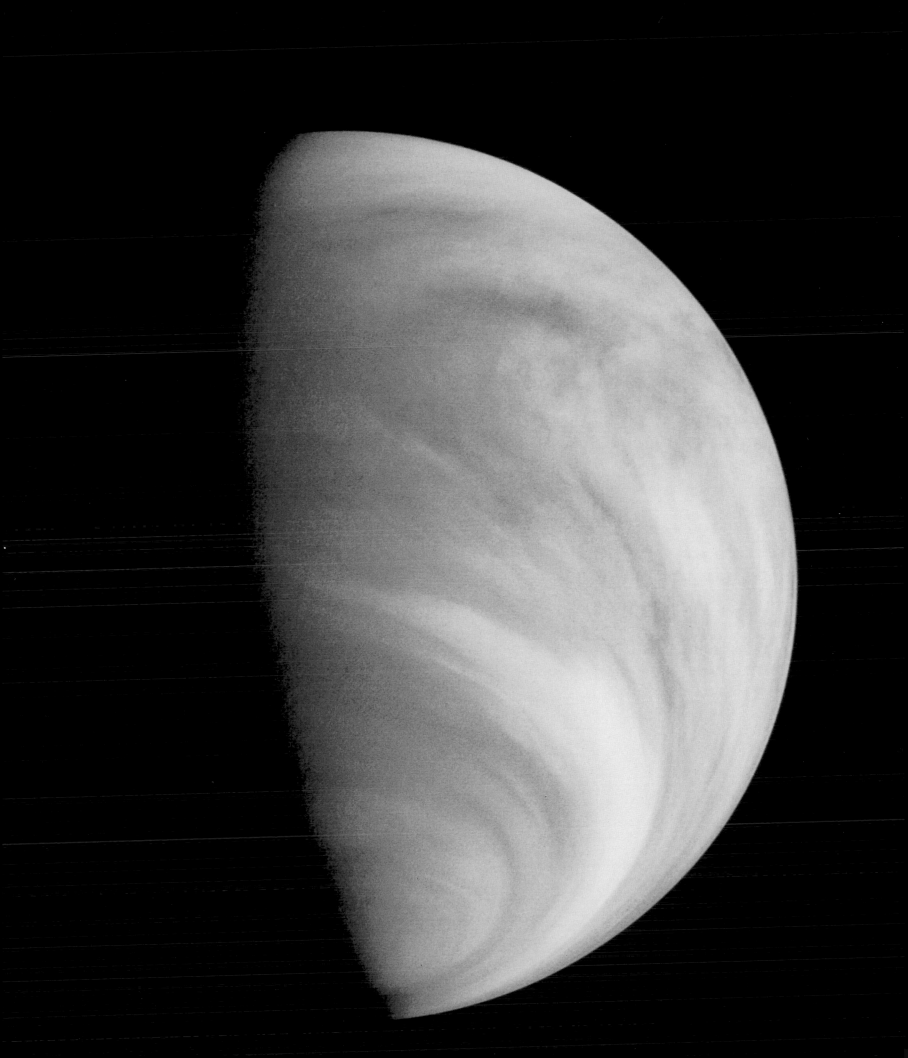

The hunt for volcanoes

To get measurements from the surface of such an inhospitable planet was no small engineering challenge, and it was the Soviets who rose to it. In 1972 *Venera 8* became the second Soviet craft to land successfully on Venus and the first to send back useful information about Venusian rock. It survived the crushing inferno for 50 minutes, and among its findings it confirmed that the rock it landed on was similar to volcanic rock on Earth. But *Venera 8* had no camera, and the Soviets couldn't wait to return with one. In 1975 *Venera 9*, along with its sister ship *Venera 10*, snapped those historic black-and-white images of the Venusian surface. Across its flat terrain were strewn what looked like slabs of volcanic lava. It was a tantalizing glimpse. Did Venus still have active volcanoes? Was it still alive, like the Earth? To answer that, however, scientists needed a global view, and they had to wait a further three years for it. This time it was courtesy of the Americans and an orbiting radar called *Pioneer Venus*.

In 1980, after two years circling above the cloud tops, *Pioneer Venus* completed the first global atlas of the planet that had hidden from us for so long. It saw a world of low-lying plains with a long highland region stretching halfway round the equator in the shape of a scorpion. To the north there was another mountain range that dwarfed the Himalayas. Among them, in far more detail than it had been possible to see with Earth-based radar, loomed the massive peak of Maxwell Montes. At first glance, Venus looked like an Earth drained of its water. But geologists immediately zeroed in on two mountains to the north with large, crater-like depressions at their peaks: were these the Venusian volcanoes we had been waiting for?

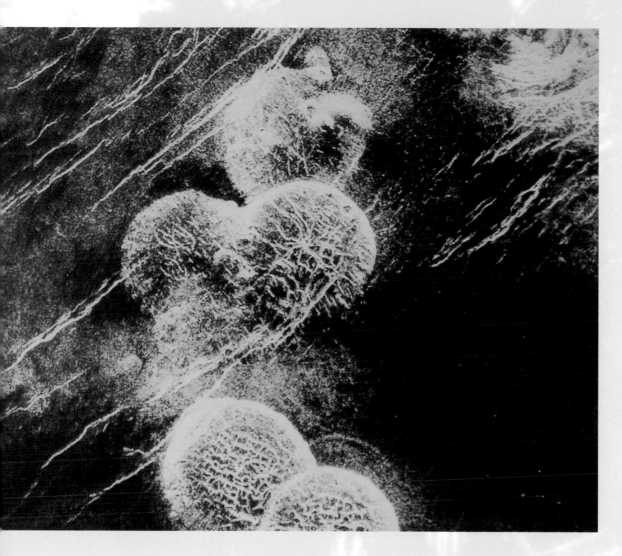

LEFT *These round hills – known as pancake domes – on Venus were imaged by* Magellan. *Each dome is about 25 kilometres across and 750 metres high.*

OPPOSITE *A giant volcano, the Aine Corona. Rising magma below the surface probably caused this 200-kilometre wide feature to bulge upwards.*

BELOW Magellan *didn't find many impact craters on Venus, but these three stood out very clearly on its radar.*

The radar map produced by *Pioneer Venus* was not much better than a child's globe, and in 1983 the Soviets decided to do something about it. They launched *Veneras 15* and *16,* a pair of improved radar mappers, to scrutinize the planet's surface with much more accuracy, seeing details as small as a kilometre across. They found Venus to be riddled with scores of large volcanoes, although none was prepared to put on a show for the orbiting duo. They also saw things that had never before been seen on Earth – strange circles of ridges up to 300 kilometres wide surrounding a central plain that seemed to have oozed with lava without ever forming the characteristic cone of a volcano. This new type of volcanic feature was named a 'corona' since it looked like a crown when viewed from above.

Although *Veneras 15* and *16* saw only the northern hemisphere, they sent back the first good aerial images of the surface of Venus. It was also the first time the Russians let the Americans share their data. A new generation of Western scientists got a taste for Venus thanks to the Soviet probes, and when the *Veneras* ceased their transmissions, they wanted more.

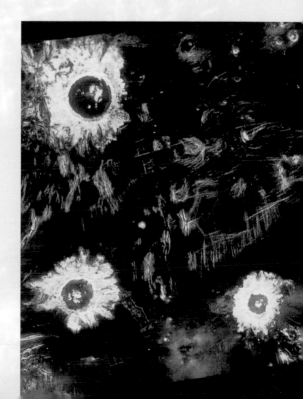

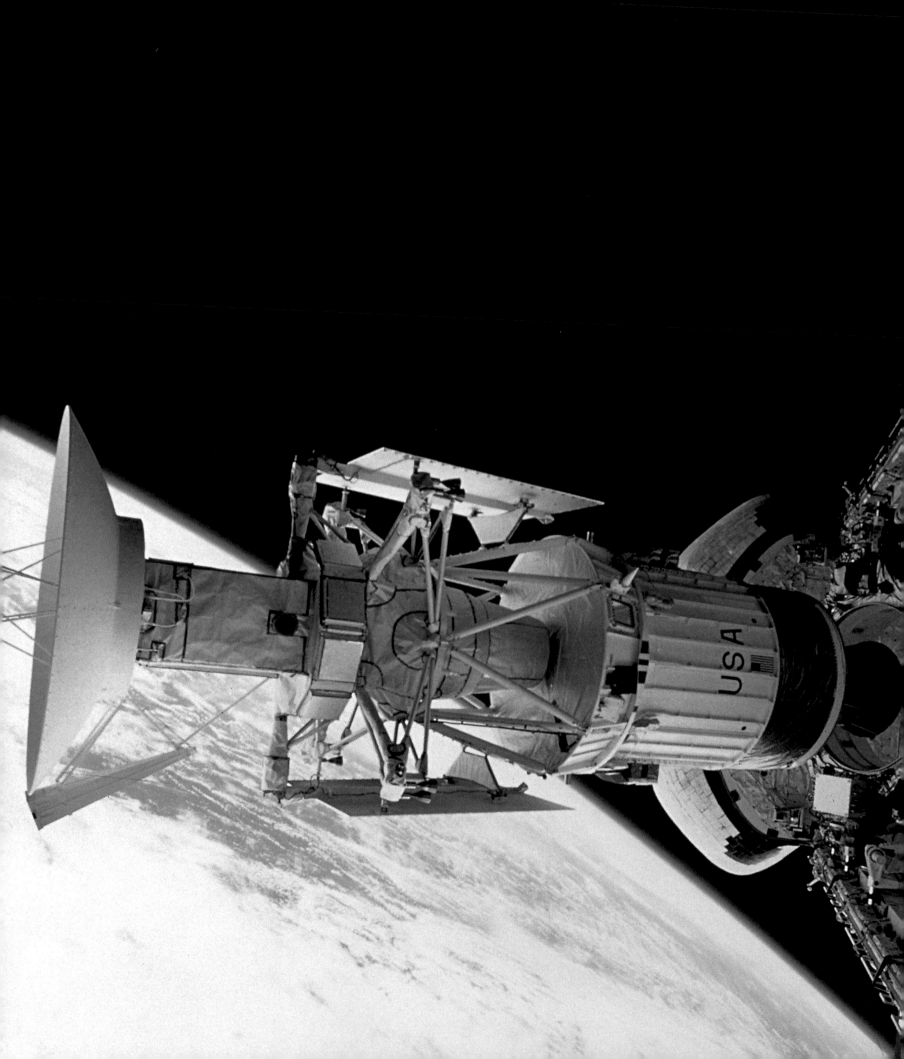

The spare-part mission

The Americans originally intended to send a grand mission to map Venus and answer any other questions a scientist might have about the planet, but the expensive project was cancelled by the Reagan administration in the early 1980s. Since some vital parts had already been built, a handful of crafty engineers realized they could still just about build a spacecraft to go to Venus and do almost as much science for much less money if they borrowed a few key parts from elsewhere. They really had to shop around. In the end, a spacecraft called *Magellan* was cobbled together from extra parts from the *Voyager*, *Ulysses* and *Viking* missions; it even included some junk from the National Air and Space Museum. The craft became known as the 'Secondhand Rose'.

On 8 August 1990 two engineers sent a signal to the vicinity of Venus to park NASA's *Magellan* probe in orbit. *Magellan* fired its retro-rockets on command and settled into an elliptical path. For the next three years its radar pierced the Venusian atmosphere and sent back a series of photographic strips covering 98 per cent of the planet in unrivalled detail: *Magellan* could see objects on the ground as small as 120 metres across. By the spring of 1991, geologists were gazing in wonder at the most detailed maps of our sister planet.

Magellan revealed Venus as a world ruled by volcanoes. It boasts no fewer than 167 volcanoes larger than 100 kilometres across. For comparison, only the Big Island of the Hawaiian chain on Earth is as large. Of smaller volcanoes, Venus boasts a staggering number – well in excess of 50,000. And there were other strange landscapes completely new to science for which geologists had to invent new names. Apart from the coronas, there were smooth raised discs, called 'pancake domes', up to 50 kilometres across. Similar domes on Earth are caused by the upwelling of pools of magma, but are not even one-hundredth as big. Most incredible of all, though, were the vast volcanic plains everywhere *Magellan* looked. At some point, lava seems to have flooded most of the planet.

There was disappointment in *Magellan*'s maps for those geologists who had hoped to prove that Venus, like the Earth, was shaped by the forces of plate tectonics. The system of volcanic ridges just wasn't there. The pattern of craters, however, suggested that the geology of Venus was up to something strange. There were only 900 impacts on the whole planet, which indicated that the surface couldn't have been around that long, and that it was, on average, a similar age to the Earth. But, strangely, the craters were spread evenly over the entire planet. The conclusion was hard to swallow: the entire surface of Venus is essentially the same age. Depending on how you calibrate the cratering clock, it is somewhere between 200 and 800 million years old.

This is a very different situation from the one we find on Earth, where it is possible to find areas of all different ages, ranging back billions of years. There is one, still controversial, explanation for the face of Venus being so uniformly young. Some geologists suggest that its solid rock crust is much better at trapping the heat of the planet, perhaps because it is so dry. Volcanoes then have a hard time bleeding lava up on to the surface, and the look of most of the planet remains unchanged for hundreds of millions of years. On Earth, water is the key lubricant which oils the process of subduction. But on Venus, over the aeons, the gradual build-up of trapped heat eventually causes the mantle to get so hot that it bursts out in one huge, planet-wide cataclysmic eruption, destroying all the existing crust in the process. According to this theory, the face of Venus today is simply a souvenir of the last time this global disaster happened.

We now have a better overall map of Venus than we do of the Earth, thanks to *Magellan*. But the probe couldn't bring scientists the news they all secretly wanted: proof that Venus is still geologically active. During the four years it spent around Venus, no volcanoes visibly erupted, no cracks opened up, no mountains moved. There was no smoking gun. The map of the planet remained unchanged. Since Venus is about the same size as our planet and blessed with a similar amount of radioactive elements, it should still be hot inside like the Earth. Perhaps its volcanic face is a sign that Venus has been better at getting rid of its internal heat and has now cooled off. But most geologists find that hard to accept. They point out that if we had spent just four years orbiting the Earth, we would have been lucky to fly directly over a volcano while it was erupting. We just haven't looked long enough; after all, we have known Venus for only a geological blinking of the eye. *Magellan* may have peeled back the clouds, but the planet is still hanging on to its secrets.

The dead planet

One night in 1877, Italian astronomer Giovanni Schiaparelli was peering at Mars from his hilltop observatory. Sketching feverishly each time he turned, blinking, from his eyepiece, Schiaparelli was in the middle of the most thorough study of the Martian surface that had ever been made. Among many of the details this dedicated astronomer noted was a prevalence of thin, faint lines which criss-crossed the surface of Mars. He described them as 'channels' or, in his native Italian, *canali*. It wasn't long before Victorian society was full of fanciful ideas of canal-building civilizations on Mars. After all, Mars had many parallels with the Earth. Although it is only half the size our world, its day lasts only 36 minutes longer than our own. It has white polar caps that, from a distance, appear similar to the Earth's. Its axis, with respect to the Sun, is also similarly tilted, so it has similar seasons to ours, although they last twice as long, as a year on Mars is 687 days. If Venus was seen as our twin, then Mars was our smaller cousin.

When the USA finally turned its rockets on Mars in late 1964, scientists had long suspected that the planet was too cold to sustain life, but Schiaparelli's canals were still scored across every map of the red planet. On 15 July 1965 those maps were about to be redrawn, when *Mariner 4* entered the most critical stage of its mission. After eight long months and hundreds of millions of kilometres, it had a mere 20 minutes to capture the first precious images of Mars as it swooped past 10,000 kilometres above its surface. It performed heroically, grabbing 21 pictures down a narrow strip of the planet from north to south before Mars drifted out of sight.

The rate of data transfer at that time was incredibly slow — eight bits of information per second (today's probes send back information about 10,000 times faster). Pictures were stored on magnetic tape and sent back over the course of the following three weeks. When the first view of another world trickled back to JPL in California, the blurred image of the edge

PAGE 86 *Precious cargo: Magellan edges out of the space shuttle's hold in 1989, beginning its 15-month journey to Venus.*

PAGE 87 *Magellan's view of Venus, stripped of its clouds.*

LEFT *Rocket to the red planet: Mariner 4 lifts off from pad 12 at Cape Canaveral, Florida, on 28 November 1964 for its eight-month journey to Mars (above).*

Space probes

1

We owe just about all we know of the other planets to a few dozen robotic space probes. Without these deep-space voyagers, our Solar System would be just a few wandering points of light in the night sky. But what does a space probe look like floating out there in the void? How does it carry out all the tasks demanded of it, and how does it send back the news from the far frontiers of its travels?

Out in the vacuum of space there is no air resistance to worry about, so real spacecraft are not the sleek and streamlined ships of science fiction: most of them have been prickly creatures, bristling with scientific equipment. The central hub is a lightweight metal frame, usually no bigger than a small car. From this, antennas, cameras, radiation probes and magnetic field sensors sprout out in all directions, sometimes on booms several metres long. And if the probe is headed for planets close to the Sun, such as Mars, Venus or Mercury, it will also unfurl solar panels to power all its gadgetry. If it is travelling to the outer Solar System, where the light of the Sun is too faint to provide any power, the heat produced by the decay of a tiny amount of nuclear fuel will keep its systems up and running for years.

Getting lost in space would be bad, so the spacecraft keeps its bearings by using sensors that look for the Sun and another point of reference –

usually a bright star down in the southern sky called Canopus. With a lock on these two bright points that lie at right angles to each other, the probe can tell which way is up and, most importantly, know where the Earth is. To spin itself around in space, every spacecraft for the last 40 years has used a disarmingly simple trick: several small bottles of pressurized nitrogen gas with nozzles stuck on the end.

A precise fix on the position of the Earth is crucial if the probe is ever going to share its discoveries with the scientists waiting back on Earth. Back in the late 1950s, NASA realized that good communication was the key to success in space travel, so it began building its Deep Space Network of radio telescope dishes around the world. Today there are sets of giant dishes in Spain, Australia and California: as the world spins, one of the dishes is always ready to pick up a signal from a distant spacecraft. They have been designed to be able to pick up the faintest radio crackle, since the signals from space are so weak. When *Voyager 2*'s antenna was radioing back from Neptune, the strength of the signal arriving at the dish on Earth was one ten-thousandth millionth of a watt: you would have to store up the energy from it for a billion, billion years to light an ordinary light-bulb for a second.

1 Magellan *heads towards Venus. Its solar panels are not yet unlocked.*

of the Martian globe against the black of space sent a ripple of excitement through the watching scientists and journalists. The relief at receiving something, anything, was tangible and the anticipation grew. The cameras were rolling and they were ready for their close-ups.

But when *Mariner 4* finally delivered, it was a crippling body blow. On frame number seven, the surface finally came into focus and the scientists saw…craters. No canals, no riverbeds, no valleys or mountains, just craters. A sinking feeling washed over mission control. Mars looked as dull as the Moon. The Earth, an active planet, destroys its craters with volcanic eruptions or shifting tectonic plates, but the presence of so many craters on Mars meant that none of that seemed to have happened. Over the next few days more detailed pictures came back, but they only rubbed salt into the wound: more craters.

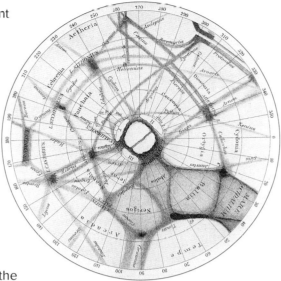

In 1969, just a few days after Neil Armstrong had set foot on the Moon, two more American craft flew by Mars. Their story was buried under the headlines of the century. Perhaps it was just as well. *Mariners 6* and *7* pretty much confirmed the findings of the first probe, sending back little more than clearer views of the wretched craters. Among the project scientists there was a feeling of anti-climax, one of the team commenting: 'We've got superb pictures, they're better than we could ever have hoped for a few years ago – but what do they show us? A dull landscape, as dead as a dodo. There's nothing much left to find.'

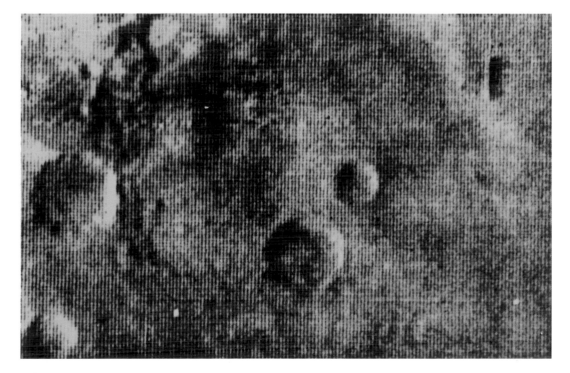

ABOVE *Giovanni Schiaparelli's drawings of Mars, made over a century ago, were still in use when* Mariner 4 *began its voyage to the red planet.*

LEFT *When the long-awaited pictures from* Mariner 4 *finally came into focus in the summer of 1965, the mission scientists could have been forgiven for thinking their probe had gone to the Moon, not Mars. All the fuzzy, black-and-white images showed were impact craters.*

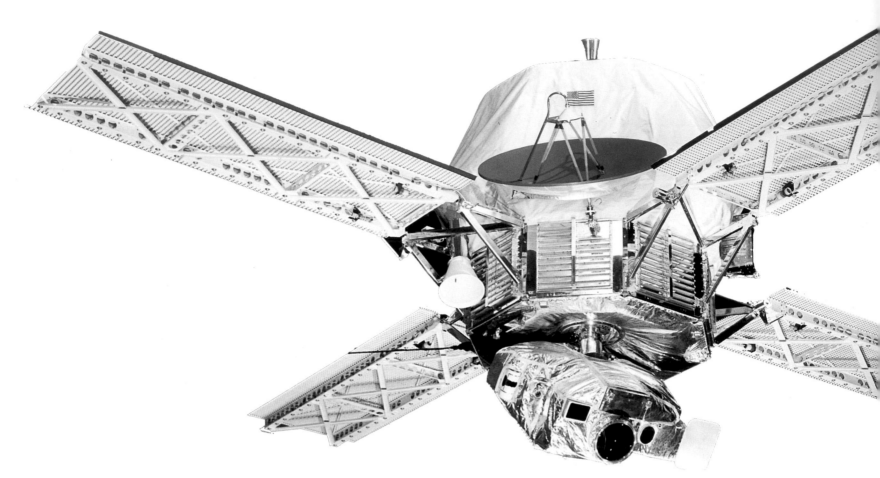

Towering infernos

ABOVE Mariner 9 *bristled with cameras and detectors below its four solar panels. The white casing above was full of rocket fuel for getting into orbit around Mars.*

OPPOSITE Mariner 9's *first view of Mars in late 1971 showed four spots poking up through a layer of dust (top right). As the dust cleared (centre) the spots were found to be volcanoes. Later images revealed dot A to be the giant volcano Olympus Mons, 900 kilometres across (bottom).*

The first three fleeting glimpses of Mars had done nothing but depress geologists. It didn't make sense that the red planet should be as barren as the Moon. It must have been a hot ball of rock when it first formed. Although smaller than the Earth, it is much bigger than the Moon, and some of its internal heat must have seeped out in the form of lava over the few billion years of its existence. Surely something on the surface should show signs of that.

On 13 November 1971 *Mariner 9* slammed on its brakes and settled into a gentle 12-hour orbit some 1,400 kilometres above the surface of Mars. *Mariner 9* was going to stay a while and map the whole planet from pole to pole. But Mars took everyone by surprise: the spacecraft was greeted by a planet-wide dust-storm. The first pictures it beamed back were completely featureless.

The year was nearly over when the dust-storms finally started to settle. As the veil gradually receded, the *Mariner 9* scientists were shocked to see four dark spots peeking up through the dust. They had expected nothing more than the rims of vast craters to break through the shroud, but these dark dots had to be many kilometres high. As the days passed, the scientists could hardly believe their eyes. The spots were quite clearly the tops of volcanoes – four of them – any one of which would dwarf the biggest on Earth.

The largest volcano corresponded to a bright spot that Earth-based astronomers had seen for years and called Nix Olympica (Snows of Olympus). It was promptly re-named Olympus Mons (Mount Olympus) and immediately took up pole position as the biggest volcano in the Solar System, rising 25 kilometres above the plain surrounding it. The other three volcanoes were only a few kilometres shorter. But incredible though they were, these giants were just the beginning of *Mariner 9*'s revelations. As the dust receded from the planet's middle, it revealed a massive canyon – a 180-kilometre-wide gash across the equator long enough to stretch right across the USA. This enormous system of gorges, in places 7 kilometres deep, was named Valles Marineris (Mariner's Valleys) in honour of the spacecraft that found it. It is the biggest feature on the planet, probably formed when the four volcanoes to its northwest bulged out the crust of Mars and ripped this gaping wound in the side of the planet.

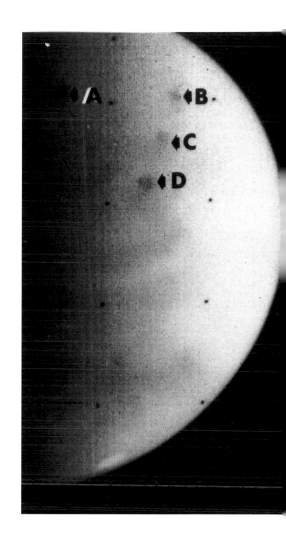

The sights that greeted the geologists now beggared belief. In contrast to the pitted surface in the southern hemisphere, the area that the earlier *Mariners* had seen so much of, most of the northern hemisphere was hardly cratered at all. Instead there were vast lava plains, areas where landslides had occurred, strange chaotic terrain, even signs of giant outwash channels where billions of gallons of water had rushed out in flash floods. There was no trace of Schiaparelli's canals, but Mars was turning out to be anything but dull. Over the next ten months, *Mariner 9* sent back over 7,000 separate images to complete the first ever atlas of this born-again planet.

But the most pressing puzzle for the new Martian geologists was to explain how a small planet like Mars could give birth to volcanoes as enormous as Olympus Mons and its three companions. It seemed strange that a planet less than half the size of ours, whose insides could never have been as hot as the Earth's, managed to erupt lava to build mountains three times the size of their biggest counterparts back here. When the scientists found the answer, it revealed something fascinating about the inner workings of Mars.

Olympus Mons grew so large because it had time; it has been gradually growing in place for 2 or 3 billion years. The four giant volcanoes on Mars are the result, like the Hawaiian Islands on Earth, of hot-spot volcanism. Below the crust, one part of the mantle is hotter than the rest (for some reason that no one really understands) and hot lava there wells up and manages to punch

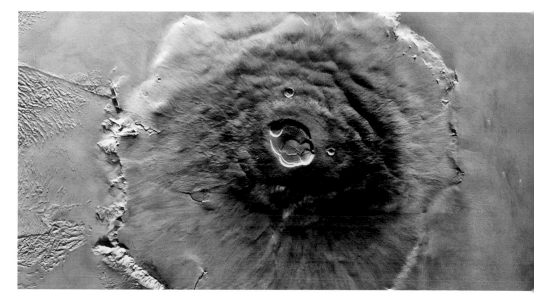

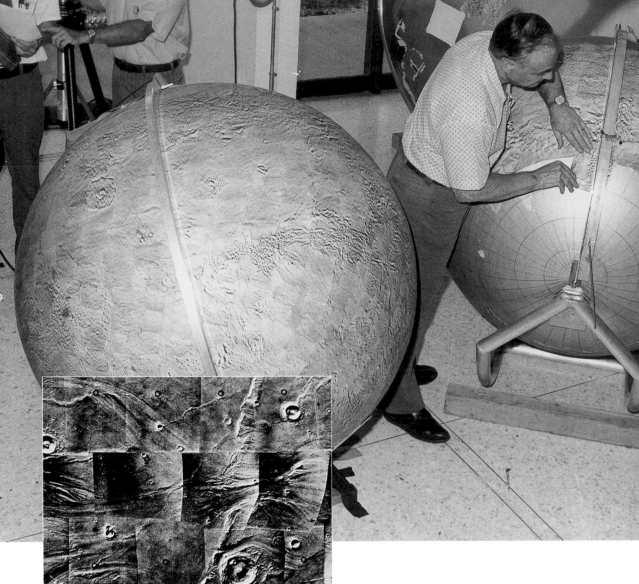

a hole through the rock above. On Earth, however, the steady drift of a tectonic plate over the hot spot means that a volcano doesn't have too long, maybe only a few million years, to form before it's shifted away from the hottest spot and begins to fade. That's why Hawaii is a chain of volcanoes rather than a single huge one. But the Martian volcanoes show no evidence of drifting. Olympus Mons has sat unmoving on top of a plume of hot magma for long enough to grow to its current immense size, every eruption adding a little more to its stature. It is almost as if Mars piled up all the Hawaiian Islands on top of one another.

It was clear that Mars had shaped its surface without the use of tectonic plates. The Martian crust is probably too cold and thick to be broken up like the Earth's, so it sits like a heavy thermal blanket around the warmer rock on the inside. Scientists were beginning to get to know our neighbouring world. *Mariner 9* had shown a new Mars to the geologists, one that had once been full of geological activity. But if Mars was back from the dead, was there any chance that it might still be alive, geologically speaking?

The   search   for   a   pulse

In 1975 the USA dispatched two more probes to the red planet. The *Viking* landings the following summer were crowning moments of triumph in NASA's unmanned space programme. Spectacular images of the surface of Mars were beamed around the world. But the landers didn't just have 'eyes', they took the 'ears' of the geologists with them too — seismic sensors to listen for Marsquakes. *Mariner 9* had showed that much of the surface of Mars was heavily cratered and therefore very ancient; however, a few areas seemed younger, particularly in the neighbourhood of the four giant volcanoes. There were suggestions that Olympus Mons might still be active. If it were, the sensors would be able to pick up the sound of its faint rumblings.

BELOW *Move over, Grand Canyon: an aerial view of part of Mars' 3,000 kilometre-wide chasm, Valles Marineris, pieced together from* Viking Orbiter *images.*

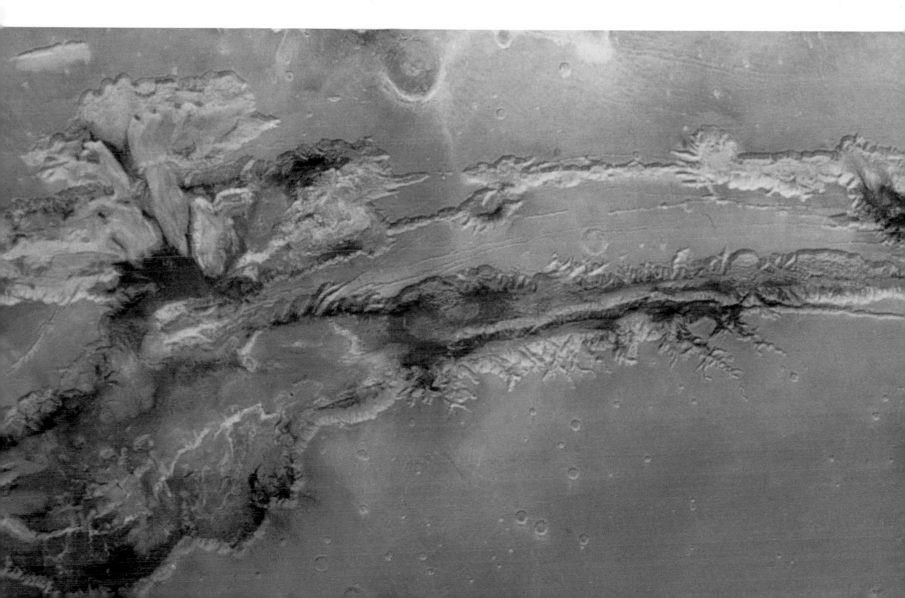

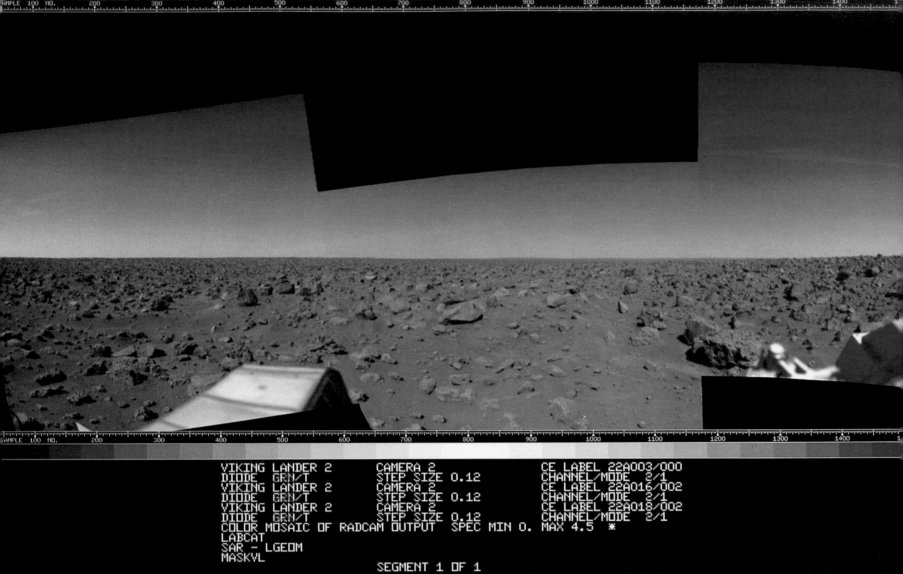

SAMPLE 100 NO.   200   300   400   500   600   700   800   900   1000   1100   1200   1300   1400   1

SAMPLE 100 NO.   200   300   400   500   600   700   800   900   1000   1100   1200   1300   1400   1

VIKING LANDER 2   CAMERA 2   CE LABEL 22A003/000
DIODE   GRN/T   STEP SIZE 0.12   CHANNEL/MODE   2/1
VIKING LANDER 2   CAMERA 2   CE LABEL 22A016/002
DIODE   GRN/T   STEP SIZE 0.12   CHANNEL/MODE   2/1
VIKING LANDER 2   CAMERA 2   CE LABEL 22A018/002
DIODE   GRN/T   STEP SIZE 0.12   CHANNEL/MODE   2/1
COLOR MOSAIC OF RADCAM OUTPUT   SPEC MIN 0. MAX 4.5   *
LABCAT
SAR - LGEOM
MASKVL
                      SEGMENT 1 OF 1
             IPL PIC ID 76/09/14/125832   WDB/L1473BX

ABOVE *Red desert: Viking 2's view of the Plains of Utopia in Mars' northern hemisphere, taken in 1976.*

Even the most successful missions have their failures, and for *Viking* it was the seismometers. *Viking 1*'s seismic detector failed to deploy, and although *Viking 2*'s got up and running, it proved a much better wind detector than anything else. Critics said the fault was not with the detector, but where it had been placed – on top of the lander rather than under its belly. Whatever the cause, *Viking* geologists never got to listen to the insides of Mars. They were left wondering if they would ever have the chance to see lava spurt from the peaks of those giant volcanoes.

NASA's *Mars Pathfinder* mission ceremonially declared the planet re-open for business on American Independence Day 1997, and marked the first time in over two decades that we had the chance to learn more about our red cousin. Pictures of the surface hit the headlines, but were strikingly similar to the vistas *Viking* saw. However, careful analysis of the radio signals *Pathfinder* beamed from Mars helped scientists probe the internal structure of the planet. The planet's daily spin injected a wobble into *Pathfinder*'s radio transmissions, and the precise nature of the wobble showed that Mars must have concentrated a lot of its mass at the centre: like the Earth, the planet must have an iron core. For the iron to have sunk to the centre, the whole planet must have been molten at one time. But are the insides of Mars still molten today?

It might be a while before we find out. An ambitious Russian probe, *Mars 96*, which was to have fired seismic sensors into Martian soil, ended up in the Pacific Ocean. In late 1997 NASA's *Mars Global Surveyor* went into orbit around Mars and began a decade of US Martian exploration by mapping the surface in incredible detail. The probe is able to see objects as small as a car. *Global Surveyor* can't see inside Mars, but it has spotted places where boulders have rolled down slopes in just the past few years. Whether this is caused by wind or by tiny Marsquakes is impossible to tell.

But there is one point in favour of continued activity on the red planet. Crater counts on the slopes of Olympus Mons show that the last eruption there was some time between 10 and 30 million years ago. That is just yesterday on geological timescales. It would be a cruel irony if it turned out that Mars had been active for 99 per cent of its history, only to shut down and die just when humans gained the power to send spacecraft to explore it.

The   first   planet

BELOW Mariner 10 *is the only spacecraft to have photographed Mercury. The blank areas are parts of the planet that have still never been seen.*

In the 1970s there was still one mysterious member of the family of worlds orbiting close to the Sun. Mercury is the hardest of the rocky planets to see from Earth since it never wanders far from the blinding glare of our star. Even during the early days of planetary exploration – a time bristling with optimism and enthusiasm – Mercury was a planet too far. The Earth is moving round the Sun at an imperceptible but breakneck speed of 30 kilometres per second. For a probe to reach Mercury it has to counteract that speed in order to fall in towards the Sun. That kind of firepower meant huge rockets, ones that were beyond the reach of even Cold War budgets. While Mars and Venus were being routinely buzzed by inquisitive probes from the Soviet Union and the USA, Mercury remained very much in the background.

But over the centuries scientists had gleaned just enough about Mercury to suspect that this quicksilver world could hold the answers to many of the key questions surrounding the formation of the rocky planets. From analysing its reflective surface, they suspected that Mercury was a world of barren and dusty landscapes, probably something like our Moon. From studying its gravitational influence on nearby Venus, they knew that it was

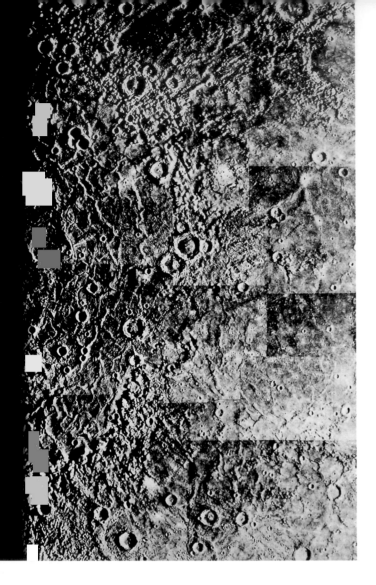

as dense as the Earth. They could peer through telescopes, they could calculate and speculate about the nature of Mercury for ever, but if they wanted hard facts, they needed a probe.

For years an idea had been discussed in the corridors at JPL, and in the early 1970s its time had come. The theory was that it was possible to change the speed and course of a craft not by expensive rocket thrusts but by using the gravitational pull of nearby planets. The idea was called 'gravity assist', and it became the cornerstone of the *Voyager* missions to the outer planets (see Chapter 4). But a mission to Mercury was to be the technology tester. If a craft were to fly by Venus on the way to Mercury, the gravity of our sister planet could change the probe's course, slowing it down enough for the Sun's gravity to pull the craft in towards Mercury.

It was just after midnight on 3 November 1973 when *Mariner 10* set off on its historic mission – two planetary encounters for the price of one. By February, *Mariner 10* had reached its first target. The probe took over 4,000 pictures of Venus and sent back a wealth of information about its atmosphere. Although *Mariner 10* followed several earlier Soviet and American probes to Venus, it was the only one designed to take pictures of the thick clouds. But the chief success at Venus was the gravity assist manoeuvre: a small error here could have meant *Mariner 10* missing Mercury by a million miles. In the end, everything went to plan and the probe fell inwards on a perfect path towards the Sun's first planet.

On 29 March 1974 the cameras of *Mariner 10* turned towards Mercury. The images, at first hazy and indistinct, steadily improved until the outline of a large crater could be seen. The first feature ever to be recorded on Mercury was named Crater Kuiper, after Gerard Kuiper, the veteran planetary scientist who had died while *Mariner* was en route. The first flyby showed Mercury to be as densely cratered as the Moon. And there was another intriguing feature – several 3-kilometre-high cliffs scattered about the planet, as if Mercury's skin had cracked at some time in the past. *Mariner 10*'s first brief encounter was quickly over but the spacecraft was now in an orbit around the Sun that took exactly two Mercurial years (176 days). This meant that every time *Mariner 10* returned to the place where it had first encountered the planet, Mercury would be there waiting for it. The craft had enough fuel for one or maybe two more meetings.

On its second encounter, *Mariner 10* sent back hundreds more pictures of the surface, confirming that Mercury had been heavily bombarded by meteorites. One impact crater, the Caloris Basin, is bigger than France and the shock-wave from the impact was so great that it created a mountain range on the other side of the planet. After its third flyby, the craft ran out of fuel and soon started to tumble out of control.

Every world has its mystery, but Mercury was the simplest planet for geologists to understand. The active life of a planet, the time during which it can mould and change its own surface, is determined by how long it can hold heat inside it. As explained in Chapter 1, since all planets were born from countless collisions of small lumps of rock orbiting the primeval Sun, all planets began life with the searing heat of those collisions trapped inside them. Ever since that day, the planets have been losing heat, and just as a jug of coffee stays warmer longer than a cup, the smaller the world, the faster it will cool. Mercury is about the same size as the Moon, and the *Apollo* rock samples told us that the Moon had been geologically dead for at least 2 or 3 billion years. Although we could not bring rocks back from Mercury, its heavily cratered face was a sign that no fresh lava had bubbled up on to its surface in a long time. There was nothing to cover up the scars of billions of years of bombardment by the Solar System's orbiting debris. Mercury is a dead world. And the unusual cliffs on its surface? Those, too, could well be the result of this little planet's molten core freezing solid. As the core cooled, it shrank, causing the overlying crust to collapse and crack like the shrivelled skin of a dried-up fruit and creating a planet-wide maze of sheer cliff faces.

Our visits to Mars, Venus and Mercury had been a slide-show of the fantastic for Earth-bound geologists. We had seen things never before imagined: volcanoes that when active would have made Earth's seem nothing more than amusing fireworks, valleys and impact craters that would swallow whole countries. But we'd seen no flowing lava, no cloud of ash, no trembling ground, no planetary pulse. Could Earth be the only planet still to have life left in it?

A different kind of engine

On 9 March 1979 at *Voyager* mission control, Linda Morabito stepped out of the small room containing her navigational computer and made an announcement that stunned the rest of the flight team. She had spent the last 45 minutes locked in a room trying to convince herself that she really was looking at something no one had ever seen before. *Voyager 1* had just flown by Jupiter, taking the first ever detailed pictures of the giant planet's clutch of moons. The craft's dash through the Jovian system had lasted just 30 hours, and most of the scientists had gone home after days working round the clock, poring over all the new data. Morabito was a member of the navigation team and it was her job to get an absolute fix on the craft's position as it rode on to Saturn. It was a tedious task that involved looking back over the *Voyager* pictures and finding the position of various stars. Often the stars were too faint and she had to over-expose the photos. It was right after she did this for one snapshot of Io, the innermost of Jupiter's large moons, that she noticed something strange. There was a bright crescent just above the surface of this moon.

At first Morabito thought it could be another moon peeking out behind Io. But she soon found out that no moon was supposed to be in that place at that time. Could it be a fault in the camera, or a strange reflection of sunlight? She checked and double-checked the image until there could be only one explanation. The strange crescent had to be part of Io: it had to be the giant plume of an active volcano shooting up debris from the surface. Morabito had witnessed through *Voyager*'s eye the first extraterrestrial eruption.

At first, geologists couldn't believe it. Finally, there was life in the rocks of another world. It was time to break out the champagne. As they celebrated this little world, it seemed

A world appears

The first images of Mars received on Earth were *Mariner 4*'s fuzzy black-and-white strips, covering only 1 per cent of the planet. It was a great achievement for 1965, but it hardly conveyed what Mars was really like. Nothing smaller than about 3 kilometres across showed up on the photos.

As the years went by, the probes became more advanced. *Mariners 6* and *7* had telescopes attached to their cameras, so they could see objects ten times smaller. Their improved communications could send back data 2,000 times faster than the earlier probes, but they captured only 20 per cent of the planet. Then, in 1972, *Mariner 9* used the same technology to make a broad-brush map of the whole of Mars. For the first time it could

be seen as a world, but there were no colours, and only objects larger than 500 metres – volcanoes, canyons, craters and ancient snaking river valleys – could be seen.

Then came the *Viking* orbiters in the late 1970s. They mapped 97 per cent of planet in glorious colour and in some places made out features as small as 20 metres across. Now hills, craters and details of the different lava flows on the sides of volcanoes appeared. Wispy early-morning clouds hugged the ochre-hued flanks of mountains. The planet's personality began to take shape – a dry, dust-strewn world covered with giant swaths of sand-dunes, laced with canyons and peppered with craters large and small.

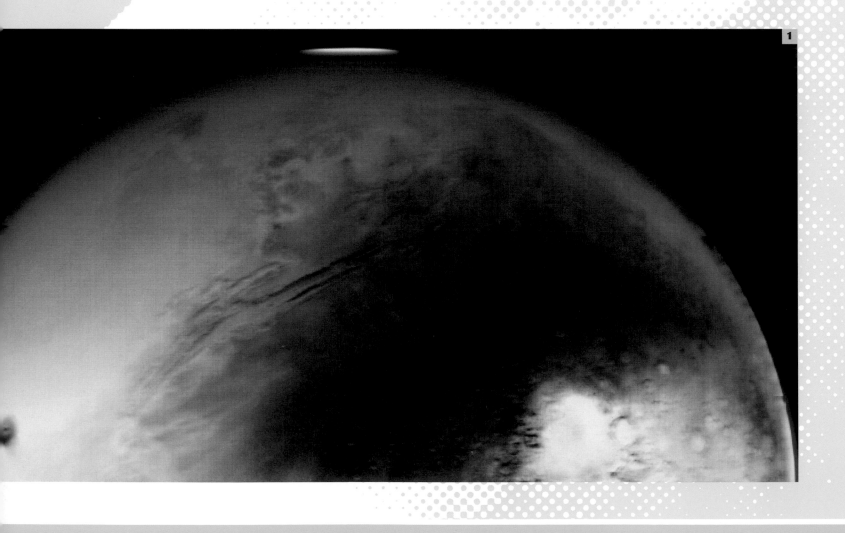

1

2

1   *This view of Mars was taken from the high point of Viking's orbit around the red planet, and the smallest visible features are many tens of kilometres across. This image is only slightly better than looking at Mars through an Earth-based telescope, but the giant swathe of the Valles Marineris canyon system can be seen cutting across its middle.*

2   *Noctis Labyrinthus: the western end of the Valles Marineris. Wispy early-morning water clouds cling to some of the canyons. This Viking view is roughly equivalent to that which* Space Shuttle *astronauts can see down on Earth when they are in orbit.*

3   *A Mars Global Surveyor image of Coprates Chasma, a small canyon inside the Valles Marineris. This picture covers an area 10 by 12 kilometres, and the smallest detectable objects are just 5 metres across, about the size of a small truck. This is what you would see if you were flying over Mars in a plane.*

More than 20 years after *Viking* it was time to crank up the magnification again. In its orbital reconnaissance mission, *Mars Global Surveyor* is currently sending back 500 million bits of digital information a day – 100 times more than *Mariner 4* returned in its whole mission. In places it can see objects as small as 2 metres across. Its spectacular pictures show sand dunes the size of those on your local beach, not the Sahara; tiny outcrops of bedrock jutting out of sandy slopes, boulders nestling in the nape of rolling valleys. As *Surveyor*'s top mission scientist Mike Malin says, looking at Mars is now like gazing down at the Earth from the window of an aeroplane. The red planet is a real place at last.

3

BELOW *The* Voyager 1 *image of*
Jupiter's moon Io. The faint crescent
is a plume of debris spurting out of
Pele, the first active extraterrestrial
volcano to be discovered.

BOTTOM *The* Galileo *orbiter caught*
another of Io's hyperactive volcanoes
– Pillan Patera – in the act in this
false-colour image from 1997.

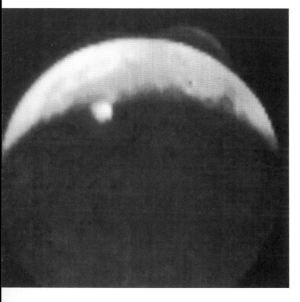

to toast them back. Closer inspection of the *Voyager 1* pictures revealed eight active volcanoes on Io. The largest one, spotted by Morabito and later given the name Pele, was throwing up a plume 280 kilometres high. In July, when *Voyager 2* showed up for the party, it found yet another volcano had kicked into life and that Pele was now quiet. The surface was a patchwork quilt of colour formed from solidified flows of liquid sulphur mingled with rock. And not a single impact crater to be seen. Io wasn't just active – it was the most geologically alive body in the Solar System. This was one heck of a moon.

Scientists were thrilled, but puzzled. Accepted ideas of planetary formation said that Io was far too small to be molten on the inside. Io is about the size of our Moon, which we know to be an inert ball of rock. What was the engine that drove Io to the point of exploding?

Tides of rock

The answer gave the geology of the Solar System a new lease of life. There turns out to be another way of melting a world on the inside: the awesome gravity of a giant gas planet. Just as our Moon raises tides on the oceans, Jupiter creates tides in the rocks of Io. Io's orbit around Jupiter is not exactly a circle, so depending on where it is in its 42-hour journey around the giant, the gravitational pull on it is slightly different. Europa, the second large moon of the Jovian system, is the culprit. It whirls around Jupiter exactly half as often as Io and regularly lines up behind it to add an extra little gravitational tug away from Jupiter, thus preventing Io from moving in a circle.

Io feels forces that make the tides in our oceans seem feeble. In fact, this little moon is close to being pulled apart. As it moves closer to or further from the giant planet, Io is alternately squashed or stretched, and the friction from these relentless forces generates enough heat to melt Io below its surface. Tidal heating, as it is now called, is a force to be

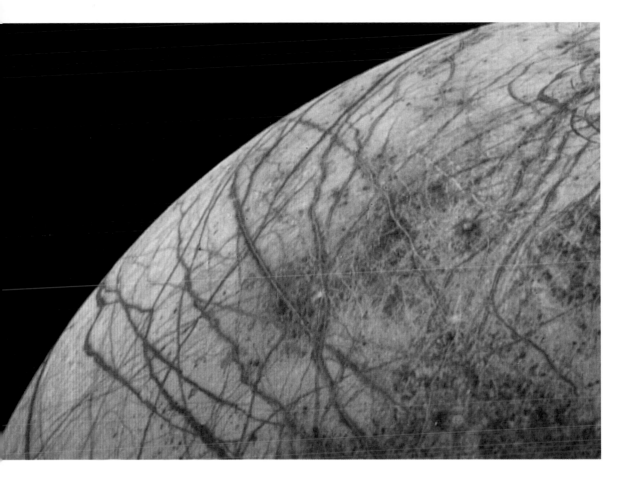

reckoned with. Oceans of liquid sulphur are thought to lie beneath Io's outer crust. Despite being a fraction of the size of the Earth, it's actually blasting more heat out into space than our planet can muster. Kilo for kilo, Io is 300 times more active than Earth.

But Io wasn't the only surprise of the *Voyager* mission. There were other perplexing worlds out there. Also at Jupiter, Ganymede, the Solar System's largest moon, appeared to have relatively few impact craters compared with its heavily pock-marked neighbour Callisto. Scientists suspected that early in its history some event melted its surface, scrubbing it clean of scars from the Solar System's violent birth.

Even stranger, from a distance of 150,000 kilometres, *Voyager 2* caught a glimpse of Europa. Covered with ice and almost completely devoid of craters, it was incredibly bright and smooth. Could tidal heating still be at play here? Even though Europa is further away from Jupiter's powerful gravity, its orbit is prevented from being circular by the pull of Ganymede, and the resulting tidal flexing could be enough to melt the ice. Some scientists even propose that there is a whole ocean of liquid water hiding beneath Europa's icy crust (see Chapter 7). The moon is scored with a global network of cracks, somewhat reminiscent of the cracks that appear in the ice-cap above the Arctic Ocean. Perhaps the daily wrenching from Jupiter was stretching and cracking an icy shell overlying a global ocean.

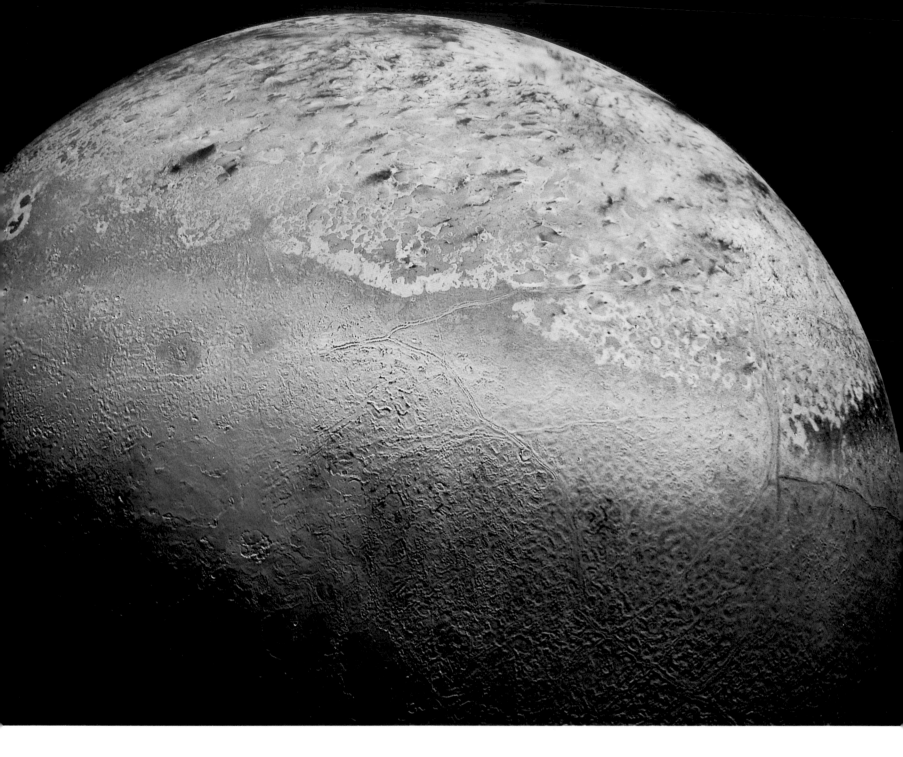

In late 1995 the *Galileo* spacecraft swooped into orbit around Jupiter to begin a four-year reconnaissance of the planet and its intriguing family of moons. It has monitored the continuing activity on Io and obtained far more detailed pictures of the surface of Europa. The best images show features as small as a few metres across. In addition, *Galileo* has shown that Europa's cracks are actually sets of ridges arranged symmetrically around a central valley. It looks like the cracks in the ice have been repeatedly opened and closed and each time liquid water has seeped up into the gap before rapidly freezing solid. *Galileo* has not yet spotted a fountain of water actually spraying upwards – a volcano fuelled by liquid ice instead of liquid rock – but all the evidence points to this being a regular event on Europa.

But back in the 1980s, as *Voyagers 1* and *2* sailed out through the dim periphery of the Solar System, they spotted other intriguing moons. At Saturn they saw that the moon Enceladus was cratered on one side, but perfectly smooth on the other, and that it was orbiting Saturn amid a ring of icy debris. Could volcanoes have spewed water from the surface of Enceladus? At Uranus *Voyager 2* found moons whose surfaces were covered with intriguing markings   water had clearly been melted there too. *Voyager* was rewriting the book on geology.

At Neptune came the sighting of the second live geological event outside the Earth. In 1989 *Voyager 2* managed just 12 pictures of Neptune's large moon Triton during its flyby. Triton is the coldest body ever encountered in the Solar System – minus 235 degrees Centigrade. But the scientists were stunned to see a series of dark streaks covering Triton's nitrogen ice-cap, which was just beginning to melt after a long Neptunian winter. After the flyby, *Voyager* geologist Larry Soderblom began piecing pairs of pictures together to make stereoscopic images of Triton's surface. He saw two dark streaks on a couple of the shots covering a similar area, and when he viewed them in stereo, the streaks suddenly jumped out at him. They weren't surface features at all – they were plumes of nitrogen gas erupting from a geyser on Triton's permafrosted skin. To top it off, Soderblom noticed one more thing. The photos were taken about 45 minutes apart and during that time the plumes doubled in length. Soderblom was seeing jets of gas and dust blowing in the flimsy atmosphere of a world 4 billion kilometres from the Sun. It was one final triumph for *Voyager 2* before it headed off into the deep black beyond the orbits of the planets.

Precious Earth

It has been nearly four decades since we first looked closely at the surface of another world. Back in the 1960s we had only just begun to understand the gargantuan forces that were continually reshaping our own planet. We went out into space in the spirit of exploration – because we simply wanted to know what was out there. Along the way we hoped we might understand more about how the Earth worked. What we discovered is that our world is actually very special. Geologists armed with the latest high-speed computers are beginning to build mathematical equivalents of worlds inside their machines. They want to see how they develop from balls of molten rock to mature planets, whether they end up like Mercury, Venus, Mars or Earth depending on their size and their distance from the Sun. They're finding that it's fairly easy to build a planet like Venus, or Mercury, or Mars. But there's something they can't seem to create in their models – tectonic plates. Maybe the Earth will cling on to its mystery the longest of all the planets.

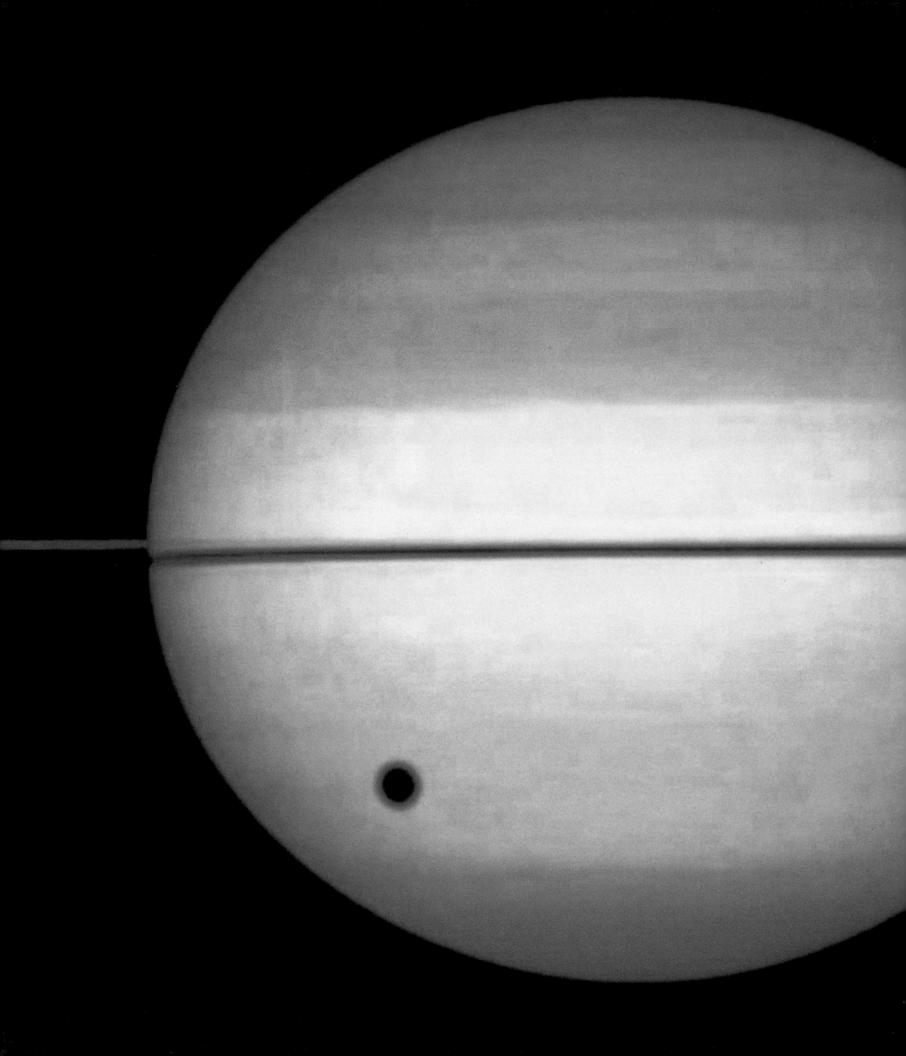

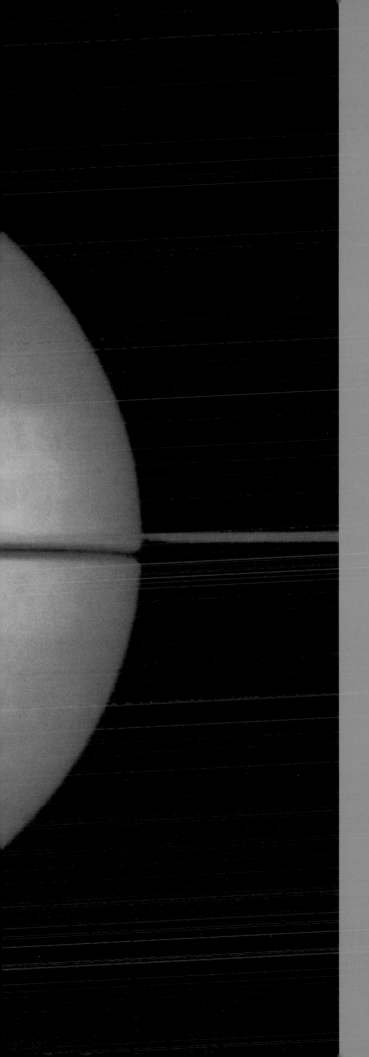

giants

IN SOME WAYS IT WAS OVER. The spacecraft performed a final pirouette, took a look at the stars to get its bearings and headed onward. The media frenzy over the first good look at Saturn and its immaculate rings had died down and only a few scientists remained at the Jet Propulsion Laboratory as *Voyager 2* receded into blackness. But as they looked at the monitors, still feeding

the latest images from the craft, they could sense something was just beginning. Saturn, picture perfect, was shrinking frame by frame as Voyager pulled away. They reflected that this view was one never before seen by human eyes: the far side of Saturn, its giant globe casting an impenetrable shadow on a broad arc of the rings. Ahead lay a journey of nearly a decade, at a blistering 40,000 kilometres per hour, to the icy worlds of Uranus and Neptune, planets so distant that scientists had no concept of how they would appear. *Voyager 2's* bearings checked out perfectly: it was on course for a celestial grand tour that would change our picture of the Solar System forever.

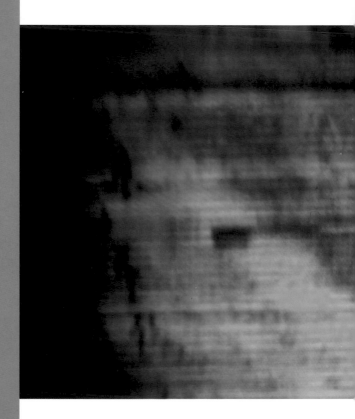

by 1972 our local planetary neighbourhood had become increasingly familiar to us. Through the latter half of the 1960s both Soviet and American craft had been skimming and probing Venus and Mars, but nothing had ventured beyond. Then, in July of that year, an American probe pushed forward into new territory. As the red disc of Mars receded behind *Pioneer 10,* there was a distant glint in its eye, a point of light larger than any star: Jupiter, the nearest and largest of the four massive worlds known collectively as the gas giants.

Jupiter is a planet big enough to swallow the Earth 1,300 times over — a world with bizarrely patterned clouds and an intriguing red spot. As we shall discover, it is also the most influential planet in the Solar System. More than half a billion kilometres beyond Jupiter is Saturn, arguably the most beautiful object in the sky, a world encompassed by spectacular rings that make it the one planet that is instantly recognizable to every person from Birmingham to Bombay. Next there is Uranus, an enigmatic ball of aquamarine, the only planet in the Solar System to roll like a barrel on its side around the Sun. Finally, Neptune, an orb of the deepest blue, lurks in a perpetual gloom, the last of the enormous worlds floating on the fringes of our horizon and our imagination. These are stormy worlds without landscapes: strange, distant, intangible.

ABOVE LEFT *For nearly 400 years, Jupiter had been seen through telescopes only as a fuzzy disc. The* Pioneer 10 *space probe changed all that when it flew past Jupiter on 3 December 1973, taking the first ever close-up pictures of this huge planet.*

ABOVE RIGHT *Tight fit:* Pioneer *tries out its nose cone for size during a check-out.*

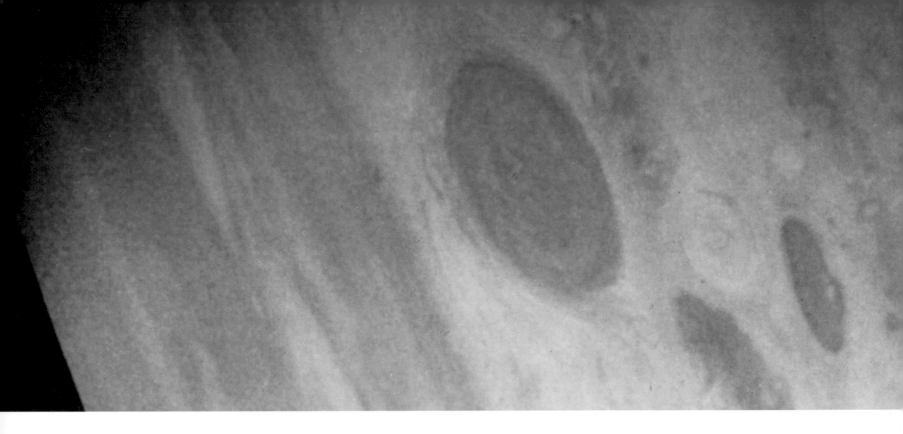

To understand these mysterious worlds would be to unravel the secrets of the Solar System itself. Millions of years older and hundreds of times larger than the rocky planets like Earth, the giants contain the primordial stuff of the Solar System. That these worlds would be revealed to us was thanks in no small part to the *Pioneer* space probes that led the way back in 1973. But before the mission scientists could even entertain the miracle of a close encounter with Jupiter, their craft had to survive a 130-million-kilometre journey through a celestial minefield known as the asteroid belt.

Into the unknown

The asteroids that form the great divide between Mars and Jupiter are spread across a band wider than the distance from the Earth to the Sun. They range in size from a speck of dust to a small country, and even though in the early 1970s astronomers had a pretty good idea of where the biggest asteroids were, no one knew how many smaller rocks might be sweeping through the belt. For seven months the scientists at NASA fretted that their precious probe would be smashed to smithereens in a collision with a rock the size of central London. Eventually, the ordeal came to an end. Miraculously, or so it seemed, *Pioneer 10* had run the gauntlet of the asteroid belt and come through intact. You could have surfed on the wave of relief at mission control. The prize for this daring manoeuvre lay across a further 280 million kilometres of empty space.

With Jupiter still 20 million kilometres away, the probe started to register unexpectedly high levels of radiation. The *Pioneer* team knew that Jupiter's large magnetic field generated more radiation than any other planet, but no one expected it to stretch so far. Ominously, the radiation increased as *Pioneer 10* closed on Jupiter like a spinning bullet, and before long the instruments had become saturated by the rising barrage of gamma rays and X-rays; its electronic instruments were going haywire. By now, several pictures of Jupiter and its moons had failed and the mission seemed to be heading for disaster. Still the radiation increased. By the time the craft was within 131,400 kilometres of the surface, it was being bombarded with 500 times the lethal radiation dose for humans; the mission was surely lost.

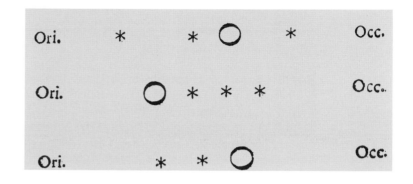

Then, just when it seemed like it was all over, the radiation levels flattened out and *Pioneer 10*'s instruments stabilized amid a storm of highly charged particles. A cheer went up in the engineering team at NASA Ames, the research centre just south of San Francisco that was running the probe. *Pioneer 10* had confirmed much of the speculation about Jupiter and shown that, if anything, scientists had been too conservative in measuring this giant world. Not only did it have a giant magnetic field, but also this planet didn't even rely on energy from the Sun. So much heat was still left inside Jupiter from the time of its formation that it was pumping out twice as much energy as it received from our star.

Having survived the intense radiation, *Pioneer 10* had work to do. Not least there were the pictures. The flickering television images that emerged back at Ames were a milestone of scientific discovery – the first glimpse of a gas giant. Bands of cloud swirled before their eyes, the ever-changing weather patterns on a gigantic, rapidly rotating ball of gas. The mysterious Great Red Spot was seen in glorious close-up for the first time. Over the centuries, many astronomers, surprised by its tenacity, had supposed it to be a solid body on Jupiter's surface. It turned out to be a storm of suitably giant proportions, big enough to engulf two Earths.

ABOVE *Galileo's drawings of Jupiter with its moons.*

BELOW *The four largest moons of Jupiter, seen from an Earth-based telescope. To the left of Jupiter are Io, Europa and Ganymede, and to the right Callisto, each moon a small world in its own right, dwarfed by the incomprehensible scale of its parent planet.*

J u p i t e r ,   t h e   w o u l d – b e   S u n

What kind of planet was it that had so nearly finished off *Pioneer*? Four centuries ago, when Italian astronomer Galileo Galilei first turned his telescope on Jupiter and saw four tiny points of light tracking back and forth along a line that crossed its equator, he realized that they were moons circling the planet. Later, renowned English scientist Isaac Newton applied his laws of gravity to the motion of the moons and got a rough idea of the mass and therefore the composition of Jupiter. Newton showed that Jupiter, though much larger and heavier than Earth, was not proportionately so. Jupiter is so large that you could comfortably fit more than 1,300 planets the size of Earth within it, but it weighs only as much as 318 Earths. The stuff of Jupiter must be much lighter than rock and iron, but what was it?

In 1903 Vesto Slipher, working at the Lowell Observatory in Flagstaff, Arizona, tried a technique called spectroscopy to find out what Jupiter was made of. He broke the sunlight reflected off

Jupiter's magnetic field

Jupiter's magnetic field

I n 1955 Bernard Burke and Kenneth Franklin, were mapping the weak radio emissions from the stars when they stumbled on a particularly large source of radiation. Over several days they recorded a signal – always at the same point in the winter sky – and decided it had to have some Earthly source. It was several weeks later that they discovered a pattern in the activity. A colleague suggested that Jupiter might be the source of the radio signal and Franklin's first reaction was to dismiss the idea as ludicrous. Nevertheless, the next day, when he checked the signals against the positions for Jupiter, he found that there was an exact match. Jupiter was bristling with radio energy – just like a star – but what could be causing it?

All sources of heat generate radio waves, even human beings. The hotter the body, the stronger the radio waves that are emitted. The energy in Jupiter's transmissions is equivalent to a billion bolts of lightning and if Jupiter itself were the source of this energy the molecules on the planet's outer layers would be blisteringly hot. Scientists knew from thermal measurements that this was not the case – in fact, quite the reverse. The cloud tops of Jupiter are a chilly 150 degrees below freezing. So what could be the source of this heat?

A young astronomer called Frank Drake made a detailed study of the wavelengths of radio energy pulsating from Jupiter and theorized that a magnetic field around Jupiter might be the source. Like the Earth, Jupiter has a magnetic field stretching out into space, only it's 20,000 times stronger. Jupiter's massive magnetic field would be continually scoured by the charged particles that flow from the Sun (see Chapter 5). The interaction between the field and the particles would 'pump up' the ionized particles to incredibly high energy levels – possibly enough to generate the 10,000 megawatts of energy detected from Earth. Drake was right in principle, but like everyone else, he had underestimated the enormity of Jupiter's magnetic field. So large is the field that it engulfs most of Jupiter's moons as well. *Pioneer 10* discovered the hard way just how far this influence extends into space. If you could see Jupiter's magnetosphere from the Earth, it would appear as large in the sky as our Sun.

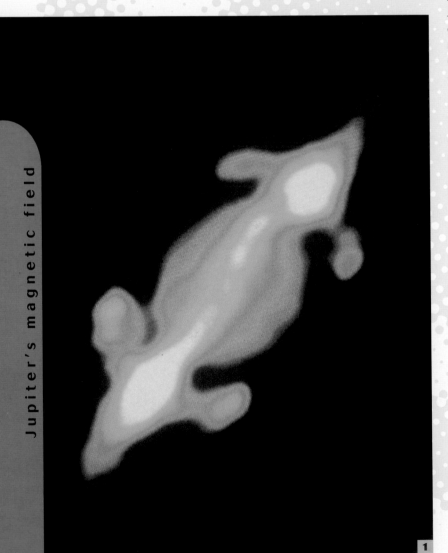

1   *Radio profiles of Jupiter. Although the temperature at the cloud tops is cool – minus 150 degrees Centigrade – the energy radiating from Jupiter's magnetic field is hotter than the Sun's surface and generates more than 500 times the lethal dose of radiation for a human being.*

Jupiter down into a rainbow, or spectrum, and hoped that certain dark lines in the spectrum (rather like a modern barcode) would reveal to him which chemical elements were floating around in Jupiter's atmosphere. But Slipher's spectra were indistinct and smudgy, quite unlike the clean, sharp lines he had hoped to be able to associate with individual elements such as carbon or oxygen. So he gave up. Some 30 years later a German chemist called Rupert Wildt came across a reference to Slipher's smudged results and realized instantly what they were. The spectral lines from Jupiter were fuzzier than expected because they didn't represent single elements: these were the spectral lines of compounds, most notably methane and ammonia. Both these chemicals contain a large proportion of hydrogen, the lightest and most abundant element in the Universe, the stuff of stars (see Chapter 5). The discovery supported a long-held theory explaining Jupiter's low density: it was made mostly of gas rather than rock, and most of it was hydrogen.

But Wildt went on to show that this gas giant was stranger yet. At Jupiter's core, Wildt calculated, the weight of hydrogen bearing down would create pressures tens of millions of times that at sea level on Earth. How would the core behave? In 1938 he announced that, towards the centre of Jupiter, the intense pressures and temperatures would cause hydrogen molecules (pairs of hydrogen atoms) to break down. The liberated hydrogen atoms would form a dense soup, which he called 'metallic hydrogen'. No one really knows what hydrogen looks like when it's a metal, but it could be something like mercury. Scientists at Lawrence Livermore National Laboratories in California have made minute quantities of metallic hydrogen for fleeting moments, but it is a substance not meant for this world and exists only at the heart of a giant planet.

The resemblance in the mix of gases within Jupiter to those in the Sun was not lost on Wildt. As described in Chapter 1, the similarity is no coincidence. As the first planet in our Solar System to form, the young Jupiter's growing gravity sucked in the gases in the solar nebula in the same way as the developing Sun. Had Jupiter managed to swallow about 40 times more gas than it did, it too would be a star and the metallic hydrogen in its core would ignite into a nuclear furnace.

In the 1950s, now at Yale, Wildt devised a general model for Jupiter that is still believed to be largely accurate. At the very centre of Jupiter, Wildt reasoned, there is in all likelihood a core of molten rock and ice ('ice' meaning solid water and ammonia and other light compounds, which are not cold but remain solid because of the intense pressures). This core is approximately 15 times the mass of the Earth and is surrounded by fluid metallic hydrogen, which is probably responsible for generating Jupiter's gigantic magnetic field as it spins. Rising further from the core, the pressure falls until the metal reverts first to a liquid made of hydrogen molecules and finally into the more familiar gaseous form. At the very surface there is a thin veneer of cloud, in proportion no more than the skin on an apple, comprising discrete layers of ice crystals containing water and ammonia.

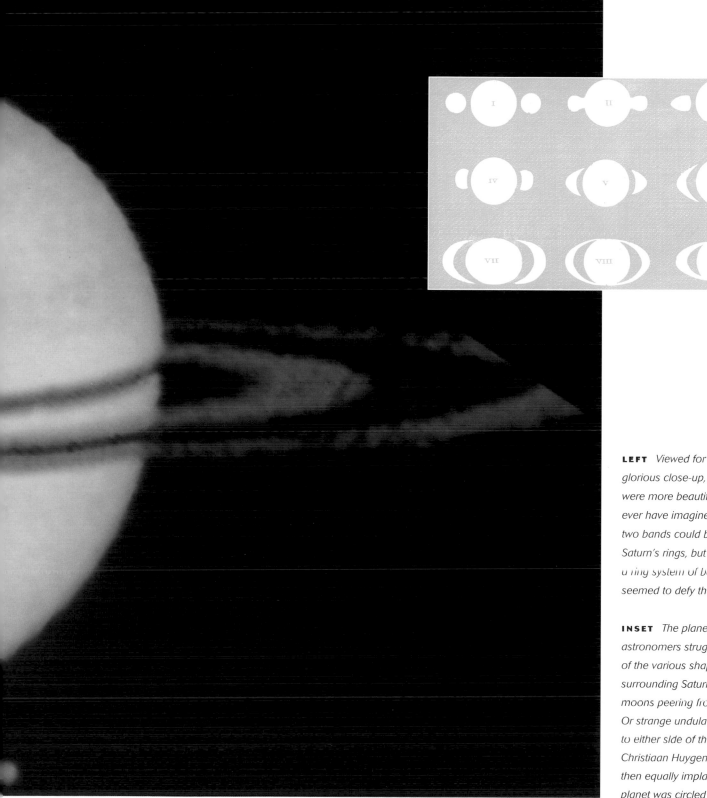

LEFT *Viewed for the first time in glorious close-up, the rings of Saturn were more beautiful than anyone could ever have imagined. From Earth, only two bands could be observed in Saturn's rings, but* Pioneer 11 *revealed a ring system of baffling complexity that seemed to defy the laws of physics.*

INSET *The planet with ears: early astronomers struggled to make sense of the various shapes they saw surrounding Saturn. Were they shy moons peering from behind the planet? Or strange undulating handles attached to either side of the equator? In 1656 Christiaan Huygens came up with the then equally implausible notion that the planet was circled by a gigantic ring.*

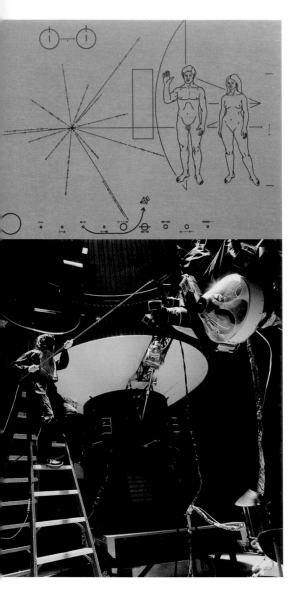

TOP *Intended as a greeting to a distant alien space-faring nation, this* Pioneer *placard gave the precise location of the Earth. Why, the feminist movement wanted to know, was it the man who was greeting the aliens?*

ABOVE *No craft has better exemplified the spirit of space exploration than the* Voyagers.

A planet with ears

A year after *Pioneer 10* left Jupiter to head off into deep space, a second craft came safely through the asteroid belt to arrive at the giant. But *Pioneer 11* wasn't just studying Jupiter, it was using the giant planet as a gravitational refuelling station, a technique first used by *Mariner 10* on its journey to Mercury via Venus. *Pioneer 11* used the pull of Jupiter like a slingshot to hurl it on towards Saturn. For five years it sped out towards the orbit of Saturn. It arrived bang on time to find the ringworld swooping into its path and on course to drag it into its massive gravitational field. One of the first discoveries made by *Pioneer 11* was that Saturn also has a magnetic field that would swamp Earth's, though it is not quite in the same league as Jupiter's. But all eyes were now on the famous rings.

Galileo may have been spot on with his ideas about the moons of Jupiter, but the rings of Saturn confused him totally. He took them to be two moons lying behind Saturn, sometimes partially revealing themselves and then sliding back behind the planet in a periodic game of peek-a-boo. His drawings of Saturn looked like the planet had ears. Fifty years later the Dutch astronomer Christiaan Huygens became the first person to realize that the 'ears' on Saturn were actually a single structure – a dazzling disc that floated around its equator. Huygens also worked out that the periodic appearance and disappearance of the 'ears' was due to Saturn being tilted to our plane of view. This tilt is probably the legacy of the last gigantic impact on Saturn during the latter stages of its formation. When the rings are exactly side-on to the Earth, they are practically invisible; as the planet moves around the Sun, its relative angle to the Earth changes and they reappear. A little later, Gian Domenico Cassini went on to discover that the rings were not solid but split into at least two separate bands. The gap between the two largest bands was named after him – the Cassini Division.

Pioneer's view of the rings was brief but dramatic. Its camera spun round and round with the spacecraft, filling in a line of picture on each turn. As the image of Saturn's girdle grew, so did the number of rings. The amount of icy dust was more than could be perceived from Earth: even the famous Cassini Division was partly filled with faint rings. *Pioneer* swooped in for its final close-up and passed clean through the ring plane. It was then wrenched off course by Saturn's powerful gravitational field and tossed out into the depths of black space like an unwanted toy.

The Grand Tour

Uranus is so far from Earth that even with the most advanced Earth-based telescopes it is impossible to make it out as anything more than a small fuzzy disc. Its remoteness had also ruled out the possibility of sending a craft to visit it – a single mission to Uranus would take 30 to 40 years and use so much rocket fuel that it would be prohibitively expensive. It's fair to say that Uranus would be an enigma to this day had it not been for a moment of inspiration by a part-time PhD student at NASA's Jet Propulsion Laboratory (JPL) in Pasadena, California.

Discovery of Uranus

From the beginning of recorded history until 13 March 1781, the Solar System comprised the Sun and the six planets from Mercury to Saturn and no more. Nobody had ever questioned this simple fact. Then, quite unexpectedly, an amateur astronomer doubled the size of the Solar System. From a terraced house in Bath, England, William Herschel gazed through his home-made telescope and discovered Uranus.

'In the quartile near Zeta Tauri…is a curious either nebulous star or perhaps a comet.' With those simple words William Herschel marked his discovery – the result of a methodical and arduous review of the entire night sky. But while tracking this curious disc over several days, something puzzled him. The object moved like a comet yet had no tail. He mentioned this strange phenomenon to friends in the Royal Society in London and soon the word spread. Within months every astronomer in Europe was tracking the comet and frantically trying to calculate its orbit. Soon

1

a consensus was reached: William Herschel had indeed discovered a seventh planet.

Herschel measured his wandering disc of light at over 50,000 kilometres wide, making it four times the diameter of Earth. He also estimated that its orbit was twice as far from the Sun as Saturn. Professional astronomers found this amateur's discovery hard to swallow. Jealous mutterings were heard in astronomical circles that Herschel's discovery had been down to luck. He was offended, replying: 'It has generally been supposed that it was a lucky accident that brought this new star to my view; this is an evident mistake. In the regular manner I examined every star in the heavens … it was that night its turn to be discovered. I had gradually perused the great Volume of the Author of Nature and was now come to the page which contained the seventh planet. Had business prevented me that evening I must have found it the next.'

Given his incredible achievement, perhaps we can forgive Herschel's immodesty. To his credit, he ignored the publicity that surrounded his discovery, and when the question arose as to what to call his planet, he rejected calls to name it 'Herschel' and suggested instead that it be dedicated to the King: Georgium Sidus – 'George's Star'. But in the end mythology prevailed and it was called Uranus, after the father of the Roman god Saturn.

1   *William Herschel (1738–1822).*
2   *Herschel's hand crafted telescope, with which he discovered Uranus in 1781.*

Back in the summer of 1964, Gary Flandro was working at JPL during his college holiday and was asked to join a group to calculate the trajectories for possible missions to Jupiter. One day, while he was refining his calculations, he noticed something that had eluded the army of seasoned astronomers at JPL. For a period around 1980, all the outer planets would line up on the same side of the Sun. Any craft that launched in 1977 or 1978 would, theoretically, be able to journey on past Jupiter not just to Saturn, but to Uranus and even more distant Neptune too. As important, the same craft could use the gravitational pull of the first gigantic planets to accelerate it on towards the next, thus reducing the craft's fuel requirements. (In fact, several years earlier another holiday PhD student called Michael Minovitch was the first to discover the alignment of the planets, but the line-up was just one of around 200 potential missions that Minovitch's computer had calculated. As a result, it had been buried under a mountain of missions which were then considered higher priority and it was quickly forgotten.) Such an alignment happens just once every 180 years and Flandro saw a chance to make his mark. The young student quickly told his boss, who was captivated by the possibilities of this rare opportunity. Soon after, JPL issued a press release announcing an 'Outer Planets Grand Tour'. The *Voyager* missions were born.

By the early 1970s, designers and engineers were already building the spaceship that would take us on a fantastic journey to the end of the Solar System. But by 1972, money for the project had dried up and NASA reduced the scope of the mission, limiting the tour once again to just Jupiter and Saturn. The imaging team leader, Brad Smith, was told by the NASA programme manager 'not to contemplate any changes to the mission whatsoever'. Mission scientist Garry Hunt also recalls being given the message loud and clear by NASA chiefs: 'We are not, repeat not, going to Uranus and Neptune.' The Grand Tour was officially off. However, unbeknown to the NASA hierarchy, the mission scientists worked around the new brief and, despite heavy budgetary constraints, secretly designed extra fuel loads into the craft so that they could travel on to Uranus and beyond. 'We'd miss Pluto,' said Hunt, 'but you have to leave something for your kids'.

The voyage begins

On 20 August 1977, around 13 years after Gary Flandro had first seen the possibility of the Grand Tour stare back at him from his pile of graphs and calculations, *Voyager 2* blasted off from Cape Canaveral. Two weeks after the launch, *Voyager 1* set off in chase. Three times heavier than the *Pioneers*, the *Voyagers* were each the size of a small house. They carried far more sophisticated instruments than their predecessors did. The craft also had souvenirs of Earth on board: among the maps, charts and messages intended for curious extraterrestrials was Chuck Berry's 1958 hit 'Johnny B. Goode'.

Voyager 1 reached Jupiter in March 1979. As soon as it entered the magnetosphere it started sending back thrilling pictures of the planet, showing colours, subtleties and details that the *Pioneer* craft had only hinted at. Jupiter's atmosphere was crackling with lightning and ghostly auroras, seething with eddies, bands of wind and spirals like hurricanes. The Great Red

Spot, just a vague whorl on the *Pioneer* images, suddenly leapt into focus. Brad Smith was dumbfounded: 'Our first glimpses of the intricate, spaghetti-like cloud structure of Jupiter with its psychedelic storm systems threw us into a state alternating between ecstasy and despair. How could we ever hope to understand this turbulent, chaotic atmosphere?'

At the same time, the Jovian moons were a revelation to the pair of *Voyagers*. Many had expected dull, cratered rocks, but Jupiter boasted a retinue of worlds as striking as the planets around the Sun. Most stunning of all was Io, a world covered in fresh layers of white, red, yellow and black patches, all different types of rock and sulphur compounds. Io was volcanically active — far more so than the Earth. As described in Chapter 3, Jupiter's gravity wrenches at Io's insides in an effect known as tidal heating, causing its volcanic activity. During its brief tour, *Voyager 1* recorded eight separate eruptions on Io. Its neighbour, Europa, was a rock just smaller than our Moon and coated with a shell of ice so smooth and lacking in craters that many scientists thought *Voyager 2*'s cameras were faulty. There must be heating on

OPPOSITE *Mighty Jupiter: at 142,980 kilometres diameter, this giant gas planet is big enough to swallow all the other planets with room to spare.*

BELOW, LEFT TO RIGHT *Io, the most volcanically active world in the entire Solar System; the dark and mysterious Callisto; the beautiful ice world, Europa — the network of fissures suggests that beneath its frozen crust lies a liquid ocean*

Europa, too, which has melted its surface. Next was Ganymede, the largest moon in the Solar System, a world of rock and ice bigger than Mercury. Ganymede had far more craters than Europa, but only at Callisto, the fourth satellite, did they find anything like what had been expected. Callisto was a cold, icy ball, the most densely cratered moon in the Solar System, a world where nothing has changed for billions of years. Although these four Galilean moons (collectively named after their discoverer) dominate the space around their planet, Jupiter was already known to have another nine moons and the *Voyagers* revealed yet another three. These smaller moons, measuring mostly tens of kilometres across, are an odd batch. Some of the most distant moons orbit the planet in a different direction from the others, suggesting that they are captured asteroids.

The diversity of Jupiter's moons made scientists wonder again about this planet. Not only is Jupiter made of the same stuff as the Sun, it also presides over a similar court of worlds. Its moons are a solar system in miniature, with the rockiest and hottest worlds closest in, and the

RIGHT *With its vastly superior cameras,* Voyager 2 *revealed details in Saturn's rings that* Pioneer 11 *could only hint at.*

BELOW *Prometheus and Pandora, the so-called 'shepherd moons'. Each less than 150 kilometres in diameter, they keep Saturn's F-ring from straying.*

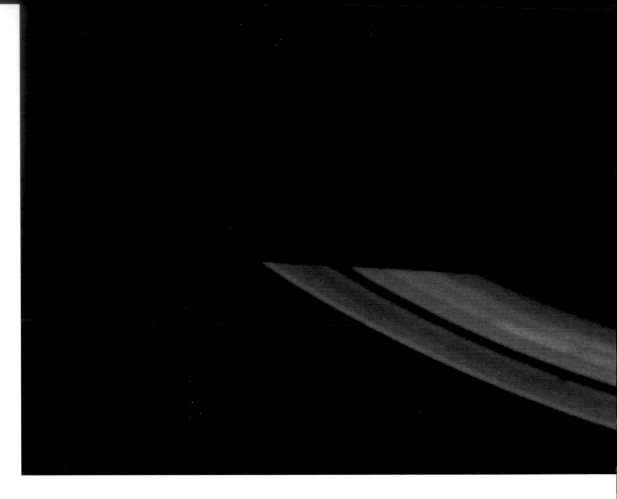

icier satellites further out. In the heat of Jupiter's formation, the young planet must have accumulated its own gas and dust cloud from which its moons formed, much as the planets had formed around the Sun.

As the probe left Jupiter, there was one last surprise. Tobias Owen and Candy Hansen, two members of the imaging team, had argued for one final parting shot. They believed that, glinting against the distant Sun, they would find a vestigial ring around Jupiter. Imaging leader Brad Smith thought they were both crazy, but estimated that he could afford to sacrifice at least one picture of the receding giant to the whims of his team. Several weeks later, Smith was sitting in his office when Candy Hansen walked in and held up a picture. He was temporarily puzzled and couldn't make out what she was showing him. Suddenly the patterns on the image fell into place and he realized that he was staring at a picture of a faint ring of debris around the great planet. Hansen and Owen had been vindicated. In fact, *Voyager* would go on to photograph similarly feeble rings around both Uranus and Neptune. But within 18 months, it would be Saturn's magnificent array of bright rings heralding *Voyager*'s next gas-giant encounter.

Rendezvous with the ringworld

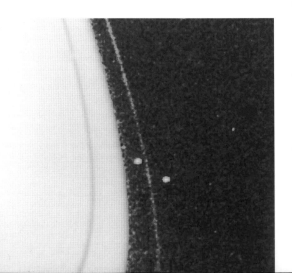

In October 1980, still about 50 million kilometres from Saturn, *Voyager 1*'s high-resolution cameras whirred into action. The first surprise was the rings themselves – they were fantastically complex structures. Under the eagle eyes of *Voyager* the rings showed far more

detail than *Pioneer 11*'s comparatively myopic sight had revealed when it passed through just over a year before. The system had been thought simple: rings of orbiting debris with gaps in between held in place by gravity. But, as it passed behind one ring, *Voyager* spotted a particularly bright star twinkling like crazy – each ring was in fact made up of literally thousands of tiny bands of debris: the Cassini Division alone contained more detail than had previously been assumed to exist in the whole ring system. Every new image showed something bizarre. The rings were not perfectly circular; one was made of two strands of material that seemed twisted around one another; there were even dark bands radiating outward across them like the spokes of a wheel.

The *Voyager* team spotted one more ring that should not have existed. It was so narrow that, theoretically, its material should have dispersed into space. Instead its edges were sharply etched against the darkness. What kept this flock of particles in such a tight formation? The answer came later that month when scientists spied two tiny moons. One was orbiting just beyond the outer edge of the ring, pushing the particles inwards; the other moon was hugging its inner edge, pushing the particles the opposite way. They quickly became known as the 'shepherd moons', gravitationally herding the ring particles into confined orbits.

As *Voyager 1* dropped through the ring plane to encounter Saturn itself, the parade of surprises kept coming. Instead of the bland atmosphere visible through telescopes, the probe surveyed a vista of swirls and storms that seemed a muted form of what had been seen at Jupiter. Below the clouds, *Voyager*'s sensitive gravity readings added weight to the suspicion that the ringed planet had a similar internal structure to its big brother: liquid and metallic hydrogen overlying a rocky core. Although it is smaller and cooler than Jupiter, Saturn is also a planet that wants to be a star.

The *Voyager* missions

Voyager 1

5 September 1977 Launched on a *Titan III* rocket from the NASA Kennedy Space Center, Cape Canaveral, Florida.

5 March 1979 Arrival at Jupiter. Flew within 277,000 kilometres of its cloud tops.

12 November 1980 Arrival at Saturn. Flew within 125,000 kilometres of its cloud tops.

Voyager 2

20 August 1977 Launched on a *Titan III* rocket from the NASA Kennedy Space Center, Cape Canaveral, Florida.

9 July 1979 Arrival at Jupiter. Flew within 651,000 kilometres of its cloud tops.

26 August 1981 Arrival at Saturn. Flew within 101,000 kilometres of its cloud tops.

24 January 1986 Arrival at Uranus. Flew within 81,500 kilometres of its cloud tops.

25 August 1989 Last stop – arrival at Neptune. Flew within 5,000 kilometres of its cloud tops.

Apart from studying the planets themselves, the *Voyagers* found 22 new moons: three at Jupiter, three at Saturn, ten at Uranus and six at Neptune.

After its encounter with Saturn, *Voyager 1* was sent hurtling northward out of the ecliptic plane – the plane in which most planets orbit the Sun. Along with *Voyager 2*, it is now on an interstellar mission, heading out of the Solar System on a search for the boundary between the Sun's influence and interstellar space. The *Voyager* spacecraft are expected to return valuable data for another two or three decades. Communications will be maintained until the *Voyagers*' nuclear power sources can no longer supply enough electrical energy to power their systems. The total cost of the *Voyager* missions to date is $865 million.

Why does Saturn have rings?

Back in the 19th century the French mathematician Edouard Roche (1820–83) discovered that at a certain distance from a giant planet a moon would be ripped apart. The difference in the gravitational forces exerted on the surface nearest and furthest away from the planet would be enough to stretch the world until it fractured and split apart. His chalk-line in space became known as the Roche limit. *Voyager* found that the entire mass of the rings was similar to the mass of Saturn's little moon, Mimas, and it seems likely that the rings formed from just such a moon. Perhaps the moon was shattered in a cataclysmic collision, or was simply knocked too close to Saturn, where it was literally pulled to pieces by Saturn's gravity. Whatever the cause of the moon's destruction, the particles are held, for the time being at least, in the planet's unrelenting grip, unable to drift away and unable to coalesce into a moon once again.

When did this happen? *Pioneer* and *Voyager* both noted that the rings are incredibly clean, unsullied by the blackening caused by constant scorching in the gusts of solar wind. Some astronomers believe that this means they are less than 50 million years old – barely 1 per cent of Saturn's life span. Others have estimated that the rings will one day disappear – either by being thrown out into space or sucked into Saturn – within 100 million years of being formed. What we see today may be just a temporary embellishment. If we could look at Saturn 1 billion years in the past, or the future, Saturn would have no rings at all, but perhaps an extra moon.

1 *False-colour view of Saturn's rings taken by* Voyager 2 *from 900,000 kilometres. You can see the mysterious dark spokes that radiate out across the rings.*

2 *The Cassini Division.* Voyager 2 *saw more detail in this so-called gap than had been thought to exist in the entire ring system.*

3 *The braided F-ring that baffles physicists to this day.*

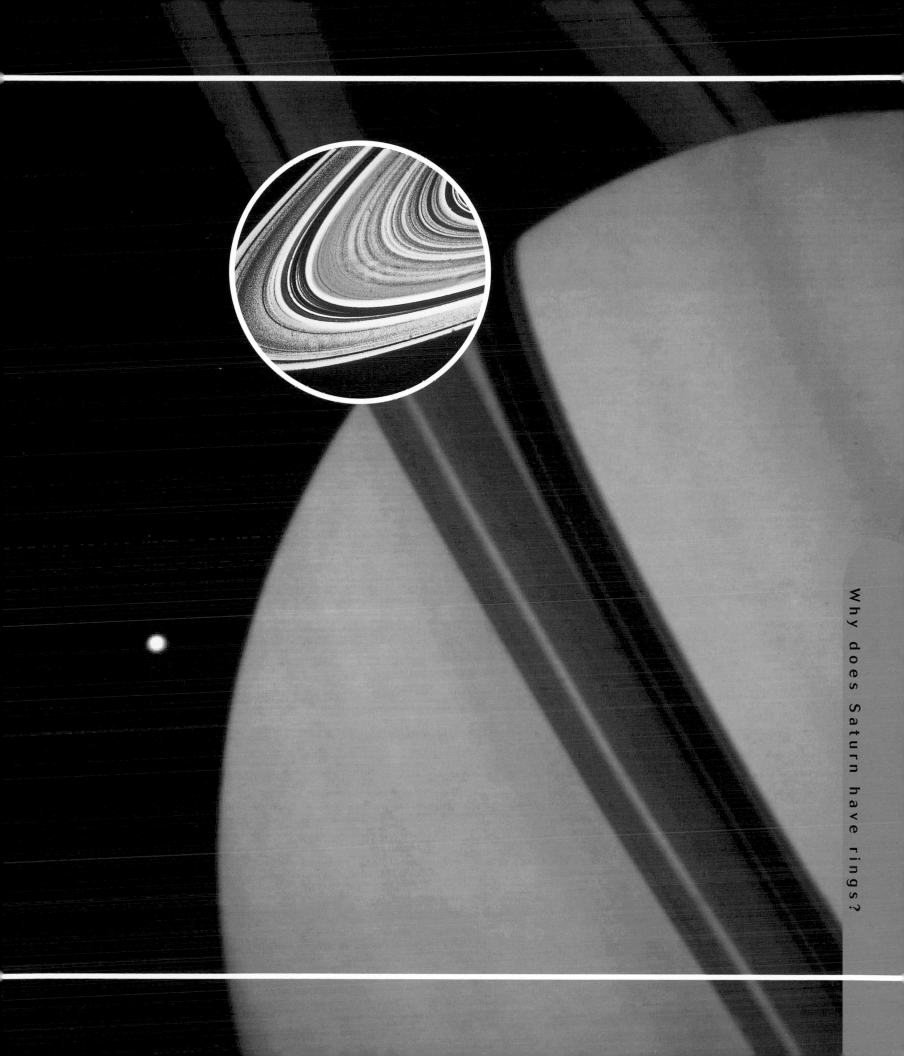

Why does Saturn have rings?

RIGHT *Saturn's largest moon, Titan, taken by* Voyager 1. *Although smaller than Mars, Titan has a foggy orange atmosphere one and a half times denser than the Earth's.*

BELOW *Iapetus, photographed by* Voyager 2 *from just over 1 million kilometres. Even from this distance,* Voyager 2's *cameras could see detail down to 21 kilometres.*

OPPOSITE TOP *Saturn's fourth largest moon, Dione, reveals its scarred surface to* Voyager 1.

OPPOSITE BOTTOM *Enceladus from 120,000 kilometres, showing details down to 2 kilometres. The ridges suggest that its icy crust has melted at some stage in the past.*

Voyager 1 also had a keen eye on the moons of Saturn, which, under its scrutiny, grew to 18 in number, the largest collection in the Solar System. Whereas Jupiter's neat array of moons replicates the distribution of planets around the Sun, Saturn's are a random bunch. Titan stands out from the crowd as the second biggest moon in the Solar System, and the only one to have an atmosphere. Its story merits separate attention and will be told in Chapters 6 and 7. Of the rest, perhaps the most enigmatic satellite is Iapetus, a sphere half snow-white and half pitch-black, probably caused by some kind of organic material or 'soot'. Then there is icy Enceladus, which, like Jupiter's Europa, shows evidence of recent melting, but Saturn's version is smooth only on one side.

Most of Saturn's satellites are small icy worlds, but they hint at a fascinating and violent history. Janus and Epimetheus are twin moons, two halves of the same fractured world which still share the same orbit. Two other moons, Tethys and Mimas, sport massive craters that cover almost their entire hemispheres. Everywhere in Saturn's system this sort of devastation is in evidence. It's a hint of the origin of those spectacular rings, which most scientists now think were formed from the remains of a shattered moon.

Onward to Uranus

Voyager 1's journey of planetary exploration ended at Saturn when it flew in for a close look at Titan before the giant's gravitational slingshot hurled it up and out of the plane of the planets. But *Voyager 2* had performed so well that NASA officials, who had by now got wind of Garry Hunt's covert programme, finally approved the Grand Tour. In August 1981 *Voyager 2* took the path through the ring plane cleared a year previously by *Pioneer 11*. It was now entering uncharted territory, guided only by the distant glimmer of Uranus that lay four years ahead; it was going to be a long journey.

The 3 billion kilometres that separate us from Uranus not only assured its late discovery but have also protected its anonymity ever since. The same Newtonian physics that had alerted us to the gaseous nature of Jupiter and Saturn told astronomers that Uranus was in all likelihood also a gassy world, but no one had any idea what it would look like.

With nearly six hours between the sending and receiving of a signal from Earth to *Voyager*, there was no chance of last-minute changes: *Voyager 2* would have to rely on earlier instructions – the probe was on its own. As it approached Uranus, the planet's south pole was

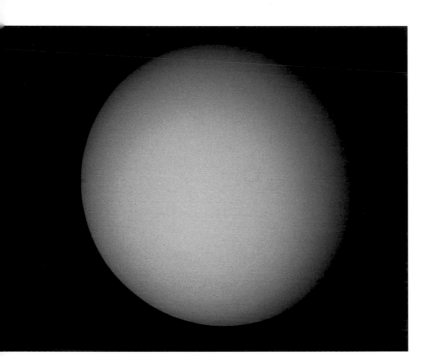

ABOVE *Uranus taken by* Voyager 2. *Scientists had hoped that a closer look would reveal some cloud detail – they were sadly disappointed.*

RIGHT *This Hubble Space Telescope image of Uranus was taken in the near infrared to enhance details of its cloud structure and its faint ring system. Because Uranus spins on its side, its moons seem to circle round it like lights on a Ferris wheel.*

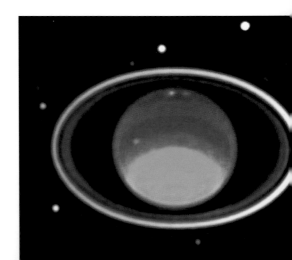

tipped towards it. *Voyager* was unable to spot any obvious reason for Uranus' 90-degree tilt and the best guess remains that, at some point in its history, Uranus was knocked flat on its back in a collision with a planet the size of Earth. One of *Voyager*'s first results at Uranus was to confirm that this world was much smaller than the earlier giants, only one-third the size of Jupiter.

And then, at last, the seventh planet unveiled its blue-green face to the camera. A layer of cold methane haze in the upper clouds, which absorbs red light, gave this planet its distinctive pale-aquamarine hue. But as the globe swelled before the camera, it didn't reveal the slightest detail; on its surface there were no bands, spots or swirling plumes to be seen. The team couldn't hide its disappointment. Andy Ingersoll, an atmospheric specialist at the California Institute of Technology, quipped that the imaging team had been renamed 'the imagining team'. Could the planet really be so bland? After Jupiter and Saturn, *Voyager* scientists had expected to find that Uranus, too, would have its own particular features. But no, the cameras revealed that Uranus was a world of uniform colour and little visible activity.

After the massive magnetic fields of Jupiter and Saturn, *Voyager* had to wait until just ten hours before its closest approach to detect a magnetic field around Uranus. It was a bizarre field too; unaligned with either pole, it dissected the planet at an angle of 60 degrees. Perhaps it was in the middle of a re-alignment, a periodic reversal of the poles. Who could tell? The mysteries just kept piling up. Skewed as it was, the magnetic field did allow the first accurate measure of the speed of rotation on Uranus. Previously estimated at 16 hours, *Voyager* announced that a day on Herschel's planet was 17 hours and 24 minutes long.

Meet the family

If the bland face of Uranus had been something of a disappointment, the planet's entourage was full of surprises. Lurking close to Uranus, *Voyager* made out the shapes of nine tiny new moons. The surfaces of the five previously known moons – Miranda, Oberon, Titania, Umbriel and Ariel – once again revealed a history of untold collisions. More surprising was that all but Umbriel showed some signs of past geological activity, melting and reforming large areas of

TOP *A great face for radio: Uranus' moon Miranda wins no prizes in the beauty pageant, but it promises some of the best mountain climbing in the Solar System.*

ABOVE *Like its cousin, Miranda, the face of Ariel tells of a chequered history. In the early days of the Solar System, huge cometary collisions would have been a daily occurrence, leaving its surface scarred for life.*

LEFT *Like Jupiter before, Uranus surprised everyone with its ring system. The icy rings are the darkest objects in the Solar System. They are shown here in false colour.*

Probes to the giants

1

Although the *Pioneers* and *Voyagers* are very different beasts, they do share some similarities. Most striking are the large radio antennae that always point back towards Earth and the long booms tipped with particle and magnetic field detectors. Both craft, travelling far beyond the scope of solar panels, are powered by small plutonium generators located away from the main body of the probes on the tips of smaller booms. On their journeys to the outer planets, the probes were steered by several small rocket boosters which made continuous tiny adjustments to their trajectory and ensured that they stayed bang on course. But there the similarities end. The *Pioneers* are much smaller craft (one-third the size of the *Voyagers*) and in order to maintain stability, the *Pioneers* gently spin as they travel through space. The *Voyagers*, on the other hand, are equipped with gyroscopes and navigate using the light of the Earth, the stars and the approaching planet.

The imaging systems on the two craft could not be more different. The *Pioneers* were fitted with a single scanning light sensor. As the craft rotated, the sensor would scan the oncoming planet and store the image, line by line, as a sequence of lighter and darker dots. Using this method, the *Pioneers* managed to send back roughly one crude colour image per hour. In comparison, the gyroscopically stabilized *Voyager* craft were fitted with high-resolution colour television cameras on top of scanning platforms, and the rate at which the data could be radioed back to Earth had improved through a combination of extra radio transmitters on the probes and better receivers on Earth. This allowed *Voyagers* to send back high-resolution colour images at the much improved rate of one every minute.

1 *NASA's* Voyager *craft, which journeyed to the giant outer planets and beyond.*

their icy skins. These alien landscapes with increasingly exotic surface features were the result of water, not rock, forming the crust. As liquid water seeped up from their hot interiors, it froze, and as the water froze, it expanded, pushing the crust apart and causing some of the most unearthly landscapes in the Solar System.

Most bizarre was the little moon Miranda, just 500 kilometres wide. *Voyager* revealed the satellite to be one of the most complex bodies in the Solar System, a collage of natural disasters. Parts of the surface were contorted into great cliffs, canyons and ridges. In one place a precipice soared up more than double the height of Mount Everest. Strangest of all were two huge, rectangular regions and another V-shaped feature, scored with ridges as if furrowed by a giant plough. The now burgeoning team of geologists on Earth put forward exotic theories to explain Miranda. Perhaps at some point in the past, she had been shattered by a collision with an asteroid and her fragments, having remained close together, gradually reassembled into a jumble of ice and rock.

Such a collision may also have been responsible for Uranus' inexplicably narrow rings. *Voyager* photographed blackened rings so slender that they should not exist. The problem was that they weren't all patrolled by the shepherd moons that keep Saturn's rings in place. Perhaps they were formed quite recently by the break-up of a small inner moon. If so, they could be short-lived, the material doomed to tumble into the planet's bland, blue-green atmosphere over the next few million years.

The Uranus encounter left scientists confused: virtually everything about the Uranian system flew in the face of scientific experience. As *Voyager* left for Neptune, Brad Smith said, 'To create a historic scenario for what *Voyager* saw at Uranus we need more miracles than any thinking person will accept.' Uranus and its violent past is still a mystery waiting to be solved.

As *Voyager* left Uranus shrinking in its wake, a point of light no bigger than a star beckoned the craft out. Billions of kilometres in the distance lay Neptune, a planet that would have escaped detection for a very long time, had Uranus not given away its position.

ABOVE *So distant is Neptune   4.5 billion kilometres from the Sun – that Voyager 2's faint radio message took over four hours to reach the Earth.*

PAGE 132 *The last lap. After a 12-year journey through the Solar System, this was the sight that finally greeted Voyager 2.*

Neptune

For the best part of 50 years after its discovery by William Herschel in 1781, astronomers tried in vain to predict an accurate orbital path for Uranus. Every time they thought they had it worked out, the planet would lurch off course like a drunk struggling to follow a white line. Increasingly desperate explanations were offered for this infuriating behaviour. Perhaps a large unseen moon was pulling Uranus off course, or a massive comet had slammed into the planet and nudged it sideways. Even the sacred laws of gravity were offered up in sacrifice when some suggested that the Newtonian Laws of Motion might end at Saturn. Despite the confusion, a more credible theory started to gain acceptance: perhaps the orbit of Uranus was being perturbed by a more distant planet lurking beyond the magnifying power of telescopes. But how to pin it down? Many people believed that the answer lay in using Newton's (still intact) Laws of Motion to perform interplanetary forensics, revealing the culprit from the gravitational fingerprints that it had left on the orbit of its nearest neighbour.

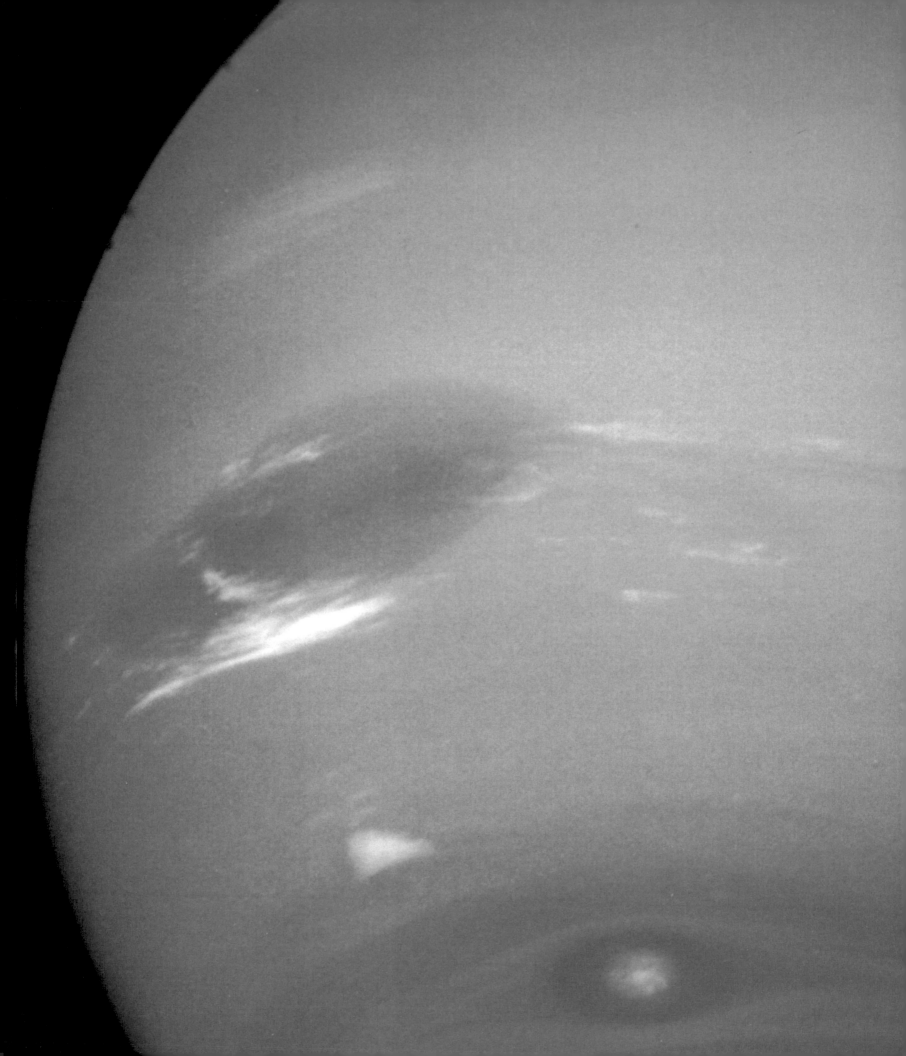

John Couch Adams, a talented undergraduate mathematician at Cambridge University, promised himself that as soon as he had finished his studies he would dedicate some time to solving the riddle of Uranus. In 1843 he emerged from his degree with double the marks of his nearest competitor and was immediately on the trail of the unknown planet. Many, including the Astronomer Royal, George Airy, thought the hunt was a mathematical dead-end. To be sure, it was a daunting task. But by 1845 Couch Adams had sent a letter to the director of the Cambridge Observatory, John Challis, which would prove any doubters wrong. In it Couch Adams not only confirmed the existence of a new planet but suggested a position for it. Had anybody bothered to check, they would have discovered that a faint spot of light was drifting within 2 degrees of his suggested location.

What followed was an embarrassment to science. Challis wrote to Airy with Couch Adams' prediction. Airy casually dismissed it as nonsense. As far as he was concerned, the

 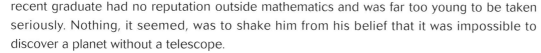 

recent graduate had no reputation outside mathematics and was far too young to be taken seriously. Nothing, it seemed, was to shake him from his belief that it was impossible to discover a planet without a telescope.

Meanwhile, in France, an altogether more established astronomer had been working on an identical solution. On 1 June 1846, nine months after the Englishman had first sent notification of his findings to the Astronomer Royal, Urbain Jean Joseph Leverrier announced the position of the planet to the Paris Academy of Science. On 25 September Leverrier received a letter from the Berlin Observatory which said simply: 'The planet whose position you have pointed out actually exists.' Although the French ferociously guarded the discovery of Neptune as their own, scientific justice eventually prevailed and both Leverrier and Couch Adams now share the credit for the planet discovered on a piece of paper.

TOP *The overlooked genius, mathematician turned astronomer John Couch Adams (1819–92).*

ABOVE RIGHT *French astronomer Urbain Jean Joseph Leverrier (1811–77).*

ABOVE LEFT *Two contemporary and highly partisan French cartoons showing Couch Adams searching for Neptune in the wrong part of the night sky, before discovering it among the pages of Leverrier's papers.*

Last stop on the Grand Tour

So distant is Neptune that in the time between its discovery in 1846 and *Voyager*'s silent arrival, the planet was still 22 years short of completing its 165-year orbit of the Sun; no one knew exactly where the planet was. When *Voyager* flight engineers managed to steer their ship to within 40 kilometres of perfection – the celestial equivalent of a hole in one from Los Angeles to London – it was, like all great achievements, a combination of skill and luck.

And so, on 24 August 1989, when *Voyager 2* arrived at Neptune after 12 years hurtling through space, the little probe was just six minutes late. Gary Flandro was at JPL with his son to witness an encounter that he could scarcely believe was happening. It had taken 25 years to get from his first slide-rule calculation to Neptune. He admits that he still has difficulty finding the words to describe the emotion he felt when he saw the first images. But Flandro doesn't take the credit: for him the real heroes were Newton and Leverrier, whose genius first took us from Earth to the furthest giant planet.

The first pictures from Neptune took 4 hours and 6 minutes to reach Earth. After the Uranus encounter, everyone feared that Neptune would be a cold, featureless world. But *Voyager*'s last encounter was to defy expectations yet again. As it drew closer to the second blue giant, bands of clouds reminiscent of those at Jupiter suddenly began to appear. It detected winds stronger than on Saturn and a great, swirling Dark Spot as large as the Earth. Neptune's high, wispy white clouds formed and dispersed in front of the cameras in just minutes: it turned out to be the most changeable planet in the Solar System.

Each day brought new discoveries. *Voyager* found another, smaller dark spot called D2 and a fast-moving white spot named Scooter. It also discovered six new moons, and a faint ring that was incomplete, clumped into arcs. Could it have formed quite recently, perhaps as a result of a comet colliding with a small moon and throwing up debris that has not yet had time to organize into a smooth ring?

The greatest surprise of all was the large moon Triton. Astronomers had always been intrigued by the fact that Triton spins round its parent planet from west to east, the opposite way to almost all the other moons in the Solar System. But that wasn't its only surprise. Whatever was expected from the coldest body in the Solar System, it wasn't geological activity. But from 38,000 kilometres above its surface, *Voyager*'s cameras saw the smoking plumes of geysers rising up thousands of metres before being blown at right angles by Triton's rarefied upper atmosphere (see Chapter 3). It was a sight truly beyond belief.

Ice giants

Despite all the superficial differences, Neptune was undeniably Uranus's twin — almost identical in size and composition. What *Voyager* found at Uranus and Neptune has forced scientists to rethink their models of the Solar System. Among the new ideas to have emerged is that instead of four similar, gassy worlds, there seem to be two distinct families of giants.

Uranus and Neptune are very different worlds from Jupiter and Saturn. Bill Hubbard at the Lunar and Planetary Laboratory in Arizona sees them as different kinds of planet altogether — ice giants, not gas giants. Beneath their freezing cloud-tops, Hubbard suggests that they are predominantly superheated and over-pressurized ices of water, ammonia and methane. They might have great oceans of water below their shrouds of gas, heated — like Earth — by the rocky cores that lie at their hearts. Their composition suggests that they formed long after Jupiter and Saturn. As young planets, Uranus and Neptune grew up in a part of the Solar System packed with rock and ice but lacking the vast clouds of hydrogen and helium that Jupiter and Saturn drew on. As explained in Chapter 1, those gases were blown away as our young Sun first bloomed into life. Both Jupiter and Saturn were already large enough to keep a firm gravitational grip on their massive gaseous mantles. Formed in a different place and in a later era, Uranus and Neptune have a character all their own.

Farewell Voyager

The encounter with Neptune was both a first look and a final farewell. On 25 August 1989 *Voyager* had run out of planets to visit. (The alignment of planets did not allow for a visit to Pluto.) Today the probe journeys on towards the boundaries of the Solar System. As we shall discover in the next chapter, *Voyager 2* and its launch mate, *Voyager 1*, are still in active service but they now have a new mission: measuring the strength and the limits of our Sun's influence (see Chapter 5). *Voyager 2* had shed new light on Jupiter and Saturn and vastly increased the paltry sum of knowledge at Uranus. At Neptune it had virtually written the book. Yet despite the heroics of the engineers and mission scientists at JPL, we have only scratched the surface of the giants. In mass and space they dominate the Solar System, dwarfing our tiny planet and its rocky neighbours. In our minds they remain bizarre, alien worlds.

Two days after the Neptune encounter was over, *Voyager*'s exhausted team members celebrated at a gala party. The guest of honour was Chuck Berry. When it came to his finale, what else could it be but 'Johnny B. Goode'? As *Voyager* sped off into the abyss, some team members couldn't quite let go of the explorer. 'I fully expect that *Voyager* will be picked up by a space-faring civilization one day,' said an emotional Rich Terrile. 'It will be placed in a museum and revered. I'm hoping that space-faring civilization will be our own. Or if not, we'll get a message in 40,000 years or so from some distant solar system and the message will say, "Send more Chuck Berry"!'

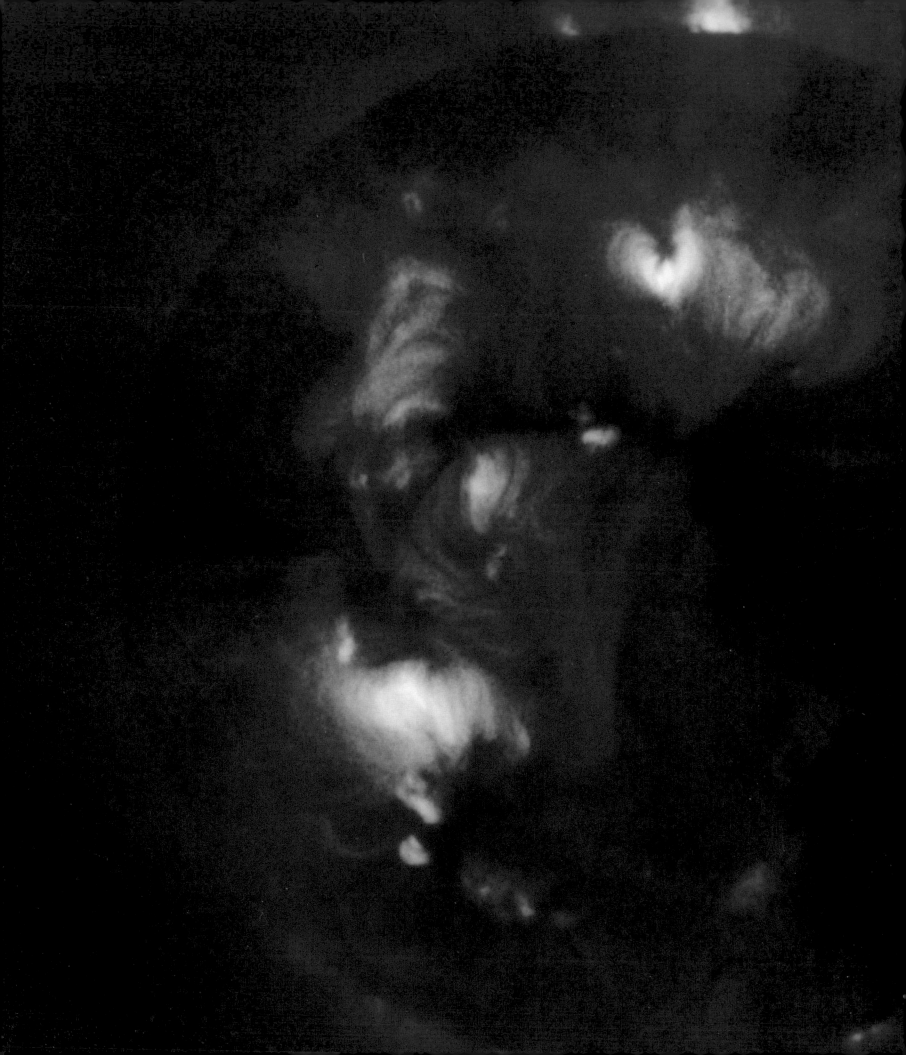

Star

#5

SEPTEMBER 1973: FOR SEVERAL
weeks now the three-man crew of
the US orbiting laboratory, *Skylab*,
has been absorbed in the most
detailed study of the Sun ever
undertaken. From their vantage point
432 kilometres above the Earth they
have been working in shifts,
recording the Sun for up to 14 hours
a day – a feat impossible from Earth.
At the beginning of their two-month
tour of duty they had left the confines
of the space station to repair a
tattered heat shield that was letting
the Sun's rays turn the lab into a life-
sapping oven. Now protected from
the Sun's deadly radiation, they were
revelling in its beauty. In the previous
week crew member Dr Owen Garriot
had become the first human being to

witness a massive eruption from the Sun's surface that had sent a ball of searing gas surging out into space at 1.5 million kilometres per hour. In his excitement, Garriot would never have believed that this was just the prologue to the most ferocious display of power ever witnessed by human eyes. 'It's like somebody's been kicking the heck out of it!' shouted fellow crew member Major Jack Lousma, as the Sun's equator came up in a rash of sunspots that sent gigantic tongues of plasma, millions of kilometres long, twisting, crackling and licking out into space. The three men sat glued to their monitors as 33 eruptions occurred in a single day. For a full week the Sun's surface writhed and spat in ways that no one could ever have predicted, building to a climax that left them breathless. 'It's the big daddy!' gasped Commander Alan Bean, as a gargantuan flare sent charged particles hurtling out into space, bombarding *Skylab* and disrupting radio communications with the Earth. They'd fixed that shield just in time.

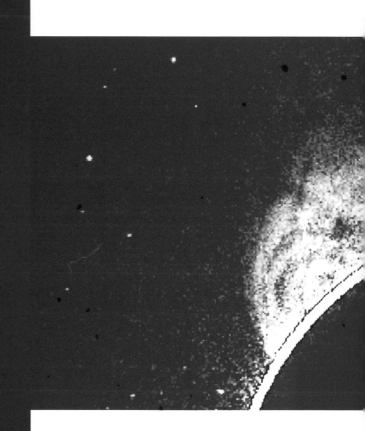

the Sun doesn't just lie at the heart of our Solar System, it *is* the Solar System. Some 99.8 per cent of all the matter between the Sun and halfway to the nearest star is contained within its titanic boiling sphere. The nine planets, the thousands of comets, the countless asteroids and the swarms of icy debris that orbit the Sun add up to little more than a speck, the tiniest fraction of material left over from its formation. As far as the Sun is concerned, the planets are insignificant grains held captive by its massive gravity and very much at its mercy.

PAGES 136–7 *This picture, taken in X-ray light by the Japanese satellite Yohkoh, reveals the invisible violence that rages on the surface of our tranquil-looking Sun.*

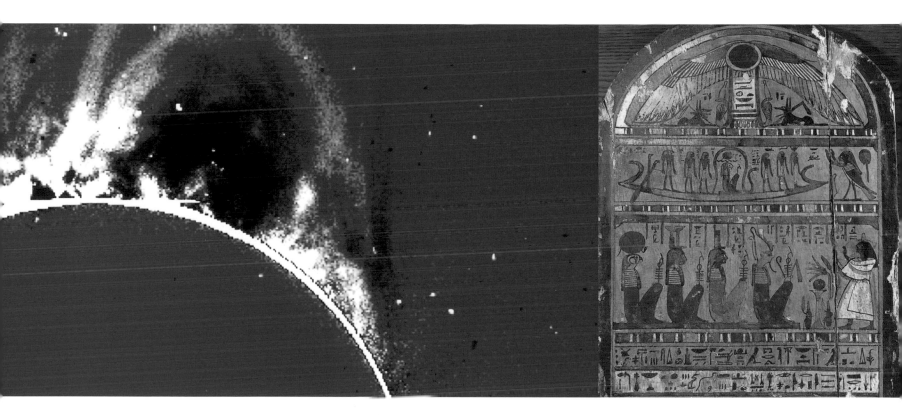

Viewed from the surface of the third planet, the blinding yellow disc that daily arcs across the sky is as familiar to us as the face of a parent, yet until very recently it has been a total enigma. One can only wonder what our earliest ancestors made of the Sun as they watched it journey from dawn to dusk. How could they explain the source of its heat? And where did it spend the night once it had dipped below the horizon, dragging warmth and light in its wake? This chapter looks at mankind's quest to understand the most important body in the planetary system. It is a journey that spans 3,000 years and takes us to the very heart of our Sun.

ABOVE LEFT *More hidden anger – solar flares, this time caught by the European SOHO satellite.*

ABOVE RIGHT *The ancient Egyptians believed that the Sun gods Ra and Khnum journeyed over their heads during the day, and at sunset were towed by barge through the underworld of Osiris where they spent the night.*

A star in our midst

To the ancient Egyptians the Sun was the god Ra, the giver of light and life. During the day, Ra was carried above their heads in a boat. At night the vessel completed its circuit back to the east on a gigantic river that ran in perpetual darkness under the Earth. The Sun has been the spark for countless mythologies, but solar science began with the Greek philosophers. One day, a flame streaked across the Athenian sky in broad daylight, heralding the arrival of a meteorite. Surely the burning lump of iron had been ejected by the Sun – or so thought the philosopher Anaxagoras in the 5th century BC. To him the Sun was a disc of glowing metal that occasionally spat chunks towards the Earth. But a century before that the Greek philosopher Xenophanes had said the Sun was a burning cloud. Here was an idea elegant and powerful enough to inspire Aristotle. The great thinker convinced his peers that the Sun was a burning ball of fire, free from imperfections, orbiting the Earth along with the Moon and the five known planets.

Most philosophers agreed that the nearest celestial body orbiting the Earth was the Moon. Beyond the Moon everything was less certain. Mercury, Venus and the Sun vied for the next position, but Jupiter and Saturn were largely accepted to be the furthest from Earth. For the next 1,900 years, a perfect ball of fire orbited the Earth from a position somewhere beyond the Moon but closer than Jupiter.

In 1543 a Polish astronomer caused a sensation when he dared to take on the accepted wisdom of Aristotle and the Church by declaring that the Sun, not the Earth, was at the centre of our Universe. While he was about it, Nicolaus Copernicus also laid out a new map for the planets, placing Mercury nearest the Sun, then Venus and the Earth, followed by

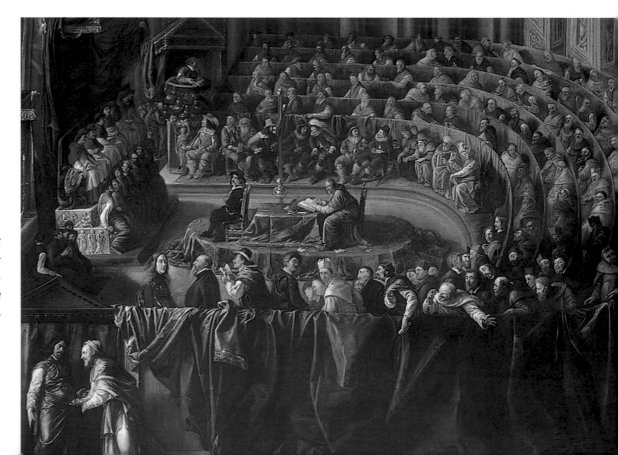

RIGHT *Galileo on trial in Rome (artist unknown, 1633). The great astronomer tried the patience of the Catholic Church when he backed Copernicus and declared that the Sun was at the centre of the Solar System.*

Mars, Jupiter and Saturn. Early in the following century, one of the world's greatest astronomers, Galileo Galilei, threw his full weight behind Copernicus' heliocentric view of our planetary system. As a result, Galileo was dragged before the Inquisition in Rome and forced to retract his view publicly. The Church might have forced Galileo into an about-face but the genie was well and truly out of the bottle. Like it or not, the Earth was no longer at the centre of the Universe – the Sun had lurched centre stage.

Some 250 years later, Father Angelo Secchi, the Director of the Vatican Observatory no less, gave the Church more food for thought. The astronomer priest was one of the first scientists to make use of the newly invented spectroscope, which allowed astronomers to deduce the elements present in a distant object simply by looking at the quality of its light. From his observatory high up in the dome of St Ignatius' Church in Rome, Secchi made a study of the chemical composition of the Sun. Some time in 1862 he decided to start a similar spectral analysis of the stars. On the very first night that he analysed a spectrum of starlight, Secchi realized that he had seen the pattern before. The spectroscopic fingerprint of the stars was, to all intents and purposes, identical to the Sun's. For over 200 years the suspicion had grown that the Sun might be a star: now Secchi could confirm it. It must have been a precious moment – the Sun and the stars, he pronounced, were members of the same family. We were living by the light of a star.

When Galileo supported Copernicus and declared that the Earth was no longer at the centre of the Universe, the Church had reacted angrily. Now their favoured astronomer announced that not even the Sun sat at the centre of the Universe; it was just one of billions of stars that stretched across the Milky Way. Astronomy had come of age and brought disquieting news. The Sun, and by association mankind, had been put firmly in its place.

By the mid-19th century the Sun might have been put into its proper context, but what about the fine details? Its brightness dazzled astronomers and protected its surface from scrutiny, so Secchi and his peers knew little more than Aristotle had gleaned centuries earlier. They could, none the less, learn a lot about the mysterious Sun, provided they chose their moments carefully.

In the shadow of the Moon

About three times every decade on average, the Moon passes directly between the Earth and the Sun offering, for a few minutes, the most beguiling and precious perspective on our star. By an incredible coincidence the Moon viewed from the Earth fits almost exactly into the solar disc. The result is that during a solar eclipse the Moon completely blocks the light from the main body of the Sun, leaving the Earth bathed in the pale ghostly light that radiates from the edge of our star. This light is called the corona and is too faint to be seen at any time other than during an eclipse. For a while, the corona was believed to be the result of sunlight being scattered by a tenuous atmosphere surrounding the Moon. But by the 1850s most scientists recognized that the corona is an extension of the Sun's surface: the atmosphere was around the Sun.

Even more mercurial than the corona is another solar secret. In the few seconds prior to totality and for a few seconds after, a red rim flashes along the edge of the Sun. Eventually, astronomers realized that this fleeting ring of light betrayed a distinct layer floating just above the Sun's surface. In normal light, this thin layer is completely invisible against the glaring face of the Sun. The Sun's visible surface was named the photosphere (*photos* being the Greek word for 'light'), and the newly discovered red translucent layer was called the chromosphere (*chromos* being the Greek word for 'colour'). During eclipses, ruby-coloured bumps had been seen clinging to the edge of the Sun; sometimes they were so large that they could be seen with the naked eye. Way back in 1231, an eagle-eyed Italian astronomer called Muratori described one of these red spots as a burning hole. And burning holes they are — tongues of flame that lick hundreds of thousands of kilometres into space before a mysterious force (or so it seemed to the Victorian solar physicists) drags their tails back into the Sun's boiling surface. Called 'prominences', these arcs of fire are among the most awesome sights in the Solar System, but until the late nineteenth century, limited technology meant that they could be seen only during eclipses. Astronomers were becoming increasingly frustrated at the few frantic seconds they had, maybe three times in every decade, to study them.

BELOW *A massive solar prominence is seen from the Sacramento Observatory in California.*

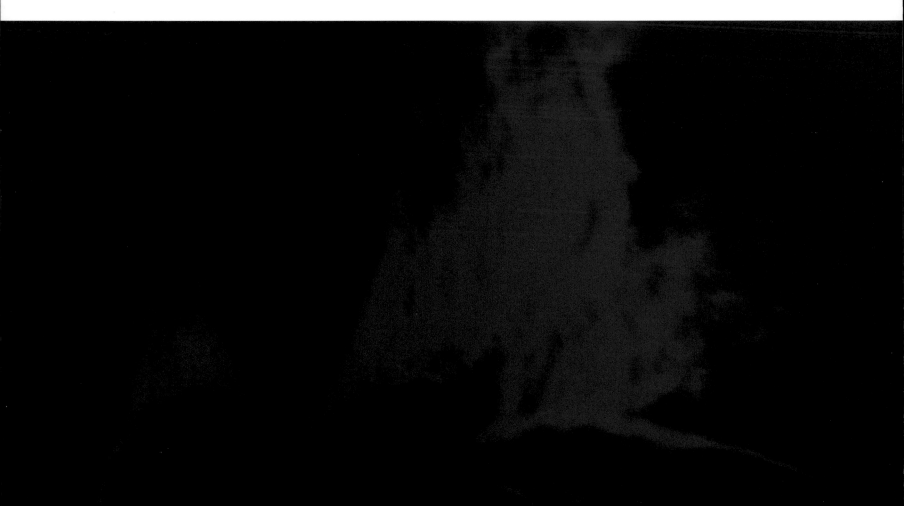

The Copernican revolution

Until the 16th century people believed that the Earth was fixed at the centre of the Universe. The idea may seem quaint now, but it was based on the self-evident truths of the physical world that surrounded them. When they watched the heavens and saw the stars drifting across the sky, they asked themselves, which is moving, the heavens or the Earth? If the Earth were moving, then an archer firing an arrow straight up into the air would surely slide from underneath the travelling arrow and it would land a safe distance from him. But in reality the arrow came straight back down on top of the archer. To astronomers the evidence was unambiguous: the Earth was unmoving and at the centre of a rotating cosmos. The idea was adopted by the Catholic Church, and the concept of the perfect heavenly system was enshrined in the Bible. The Western Universe became the work of an infallible Christian God.

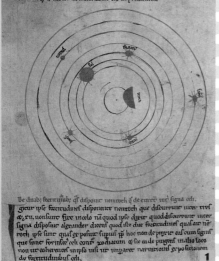

Over the next 1,000 years, however, it became harder for astronomers to reconcile that view with what they actually observed in the sky: the planets' motions were far more complicated than they should have been in a simple, Earth-centred Universe. In 1543, after a lifetime's frustration, Polish astronomer Nicolaus Copernicus' book *De Revolutionibus Orbium Coelestium* (*On the Revolutions of the Celestial Spheres*) was published, which dared to point out that if the Sun was made the focus of the planets' movement instead of the Earth, the observed motions of the planets made perfect sense. The Sun, argued Copernicus, not the Earth, was at the centre of the Solar System. *De Revolutionibus* was published in the year of Copernicus' death, and to save him posthumous ignominy his publisher inserted a disclaimer at the front of his book. He wrote that the arguments within were to be viewed merely as a mathematical device and that the assumption of the Sun's central position was not to be taken literally. The Church took this unsolicited amendment at face value and Copernicus escaped charges of blasphemy.

Nevertheless, a growing number of astronomers privately subscribed to the Copernican revolution and a new cosmic order was established. It took Galileo Galilei, fortified with the courage of his own telescopic observations in the early 1600s, to finally denounce the nonsense of an Earth-centred Universe. The Church saw this as the first direct attack on the works of Aristotle and the Bible, and after a visit from the Inquisition, Galileo publicly retracted his views and lived the rest of his life under house arrest.

1 This 15th-century illustration, from an illuminated treatise on astronomy, shows the Universe as it was then thought to be, with the Earth at the centre, orbited by the Sun and the other known planets.
2 A diagram showing the 'Copernican system', named after Nicolaus Copernicus (1473–1543), the Polish astronomer who first proposed that the Sun, not the Earth, was at the centre of the Universe.

In 1869 two astronomers, the Frenchman Jules Janssen and the Englishman James Norman Lockyer, watched an eclipse and had the same thought. Might it be possible to observe the patterns of light at the edge of the Sun without having to wait for a total eclipse? Simultaneously, the two men adapted their spectroscopes to filter out the visible wavelengths of light emanating from the body of the Sun, allowing them to see, in narrow strips, the ruby-red light flickering just above its surface. These agitated displays were suddenly observable all year round.

Once the spectroscope had filtered the glaring disc of the Sun from view, other intriguing 'landscapes' became apparent in the chromosphere. Father Secchi was one of the first to observe what he called *spicules*, meaning 'spears' or 'stalks'. These tenuous shafts of gas rush up to 15,000 kilometres high, looking for all the world like fields of burning grass blowing in the wind. About 2,000 kilometres wide, each blade fades in about ten minutes as new shoots thrust upwards to replace them. Secchi called this display 'prairie grass' and noticed that it tended to occur in ordered clumps, which, for obvious reasons, were named 'hedgerows'. Above the photosphere, the chromosphere is also dotted with flat areas that look like sandy beaches. It was this appearance that prompted French astronomers to call them '*plages*'.

Towards the end of the 19th century, astronomers were able to photograph the Sun in its entirety in differing wavelengths of light. A young solar physicist named George Ellery Hale

ABOVE *Just moments before a total eclipse, the Sun's light is all but blocked out, producing the dazzling effect known as the 'diamond ring'.*

RIGHT *Vienna in the shadow of the Moon, as seen by 19th-century artist Balthasar Wigand. The lure of a total solar eclipse has always proved irresistible.*

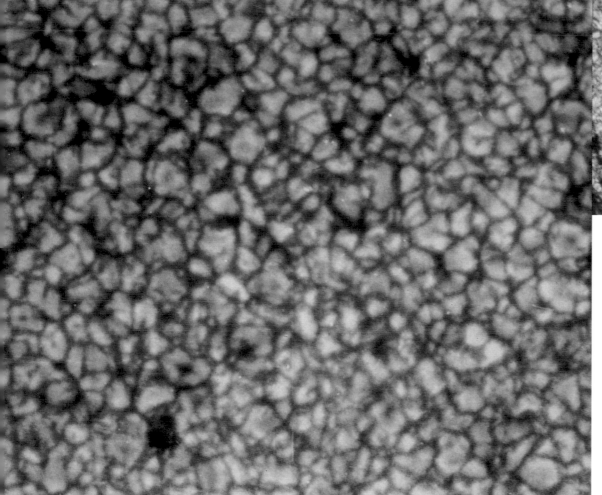

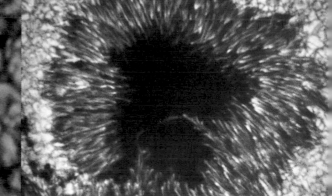

invented what he called a spectrohelioscope, and soon astronomers such as Janssen were perfecting images of Secchi's spicules and hedgerows, capturing them for eternity. Some of the most intriguing images were of the granular surface of the Sun. It seemed to be covered in millions of individual cells of rising gas. Modern measurements show that these bubbles are commonly about 1,500 kilometres in diameter. They surge up from within the Sun at speeds approaching 3,600 kilometres per hour, before cooling at their edges and sinking back down into the interior. Known as 'granules', these cells smother the star from pole to pole and together rise and fall in waves like a stormy sea. Scientists were starting to come to terms with the complexity of the Sun's energetic surface, but the processes deep inside it, which fuelled these spectacular displays, were as much of an enigma as ever.

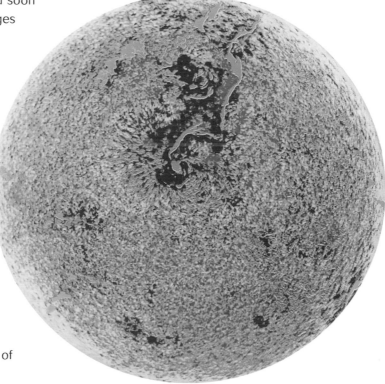

The fire within

Every person ever born has understood instinctively the importance of the Sun's energy to life on Earth. We feel its heat on our faces and know how cold the world gets when it disappears for the night. Moving beyond instinct, science has quantified the amount of energy falling from the Sun as equivalent to burning 100 litres of paraffin for every square metre of the Earth every year. But the Earth is little more than a mote of dust floating in the Sun's floodlights and collects only the merest fraction of the energy being pumped out by our star. Where does this torrent of energy come from?

Two and a half thousand years after Anaxagoras suggested that the Sun was a disc of burning metal, the best guess that scientists could make about the source of the Sun's heat was that it was a ball of burning coal. But the Victorian astronomers' ever-more sophisticated use of spectroscopy was revealing that, of all the elements that existed within the Sun, the most abundant wasn't iron, or carbon, or silicon, but hydrogen. Three centuries earlier, Isaac Newton had used his newfangled laws of gravity to show that the Sun was made of something much less dense than rock: the observations were finally bearing that out. The Sun was a ball of gas.

By the mid-19th century, Hermann von Helmholtz, a giant in the physics of his time, had reasoned that the source of the Sun's energy could not be simple combustion. When fuel burns, it reduces, like coal to soot. If a fuel were simply burning at the heart of the Sun, our star would be shrinking – but it wasn't. Putting together the known facts about solar composition, Helmholtz deduced the first scientifically plausible explanation for the Sun's heat: it was being generated by the slow contraction of its gaseous mantle. By Helmholtz's estimate, the Sun could have been producing energy in this way for 20 million years, and would continue to do so for a further 17 million before it collapsed. But Helmholtz's theory was soon in trouble. By the start of the 20th century, radioactive dating of rocks suggested that the Earth was far older than had been previously thought – at least 1.5 billion years older than the 10,000 years they had previously believed it to be (the actual age is about 4.5 billion years). According to Helmholtz's theory, the contracting Sun could be only a fraction of that age. Clearly, his theory was wrong. In fact, the first clues to the source of the Sun's heat had been discovered 50 years earlier.

In 1868 came a decisive moment. Coincidentally, it also involved the simultaneous and unconnected genius of both Norman Lockyer and Jules Janssen. The two astronomers were studying the chemical composition of prominences during a solar eclipse, when, among the familiar pattern of spectroscopic lines emanating from the Sun's surface, an entirely new line suddenly stood out to them. The position of the line was in the yellow region of the spectrum and had no known Earthly counterpart. They both realized it was the mark of a new chemical element, and Lockyer named it after the Greek sun god, Helios. Helium, found to be plentiful on the Sun, but almost non-existent on Earth, quickly became the Sun's biggest mystery: it would turn out to hold the key to the source of solar power.

ABOVE *Sir Norman Lockyer (1836–1920), who discovered that the Sun was brimming with a new element – helium. It was the key to understanding the way the Sun generated its heat.*

Sunlight's 10-million-year journey

I f you could, by some miracle, lift the lid on the pressure cooker and look into the heart of the Sun, you would see nothing at all. Of all the energy generated there, none of it is visible to the human eye: the heart of the Sun is blacker than pitch. The by-product of the nuclear reaction that converts hydrogen to helium at the centre of the Sun is energy in the form of gamma rays, high-energy radiation. Under normal conditions these gamma rays would take just 2.5 seconds to reach the Sun's surface and stream off into space. But the density of the Sun's 15-million-degree core is such that before the gamma rays travel even 1 centimetre, they collide with a subatomic particle and degrade into lower-energy X-rays. The X-rays have no easier passage than the gamma rays, bouncing backwards and forwards for what approximates to an eternity.

Eventually the individual X-rays make it to the outer fringes of the nuclear core, where they enter the next layer of the Sun: the radiation zone. From here on, as the Sun cools to just 5 million degrees and becomes much less dense, the nuclear reactions stop and the X-rays make their way relatively swiftly to the next layer, known as the convection zone. Here the temperature drops again to a mere 1 million degrees, but again the passage of the X-rays is blocked. Instead, they are carried to the surface in gigantic tunnels of scorching gas. Travelling at over 300 kilometres per hour, the gas eventually wells up and out of the Sun, appearing as the pattern of tightly packed granules that cover the surface. Now the X-rays undergo other mutations, losing more energy and becoming first ultraviolet and finally, at the surface, sunlight. The journey from the core to the surface that would take an unfettered gamma ray just 2.5 seconds, has taken around 10 million years. After such a gruelling journey, the newly liberated sunlight takes just eight minutes to reach the Earth,

40 minutes to reach Jupiter and in just over seven hours it has left the Solar System behind. Every shimmering pocket of starlight scattered across the night sky started its journey in a similar way before travelling hundreds, thousands, even millions of years to reach the Earth.

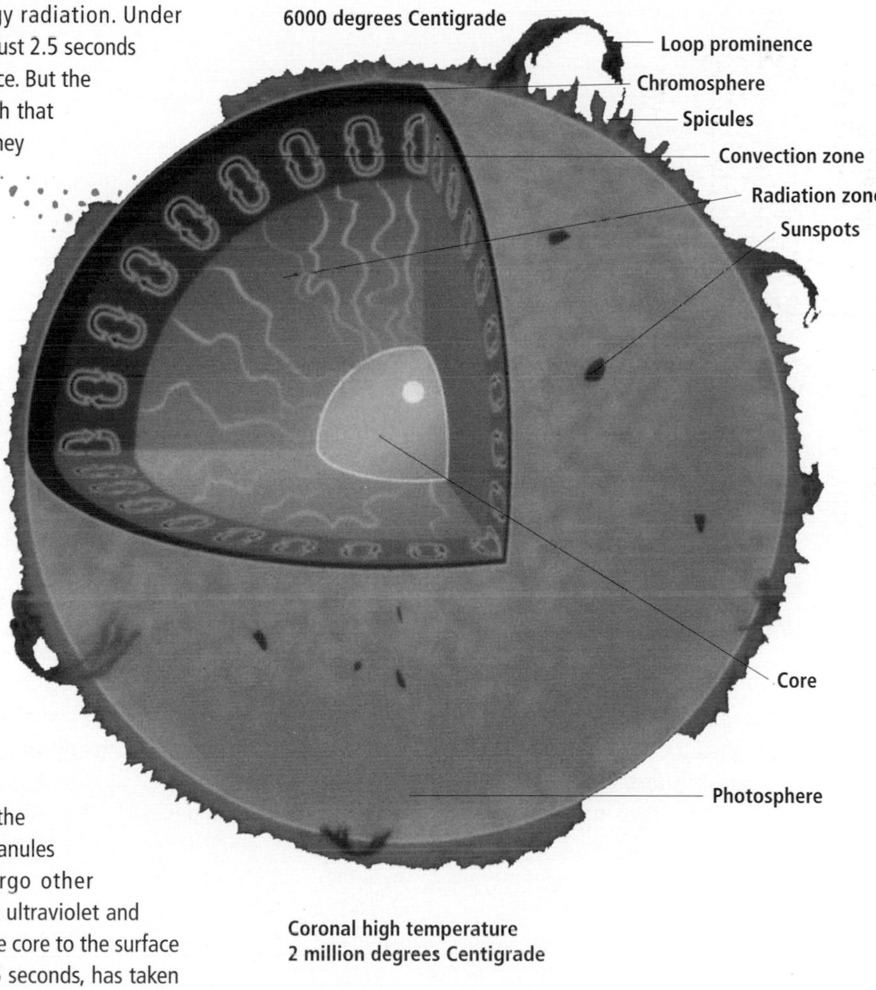

6000 degrees Centigrade

Loop prominence
Chromosphere
Spicules
Convection zone
Radiation zone
Sunspots

Core

Photosphere

**Coronal high temperature
2 million degrees Centigrade**

A common or garden star

Stars, like people, come in all shapes and sizes, and while our Sun may be of unique importance to us, among the hundreds of billions of stars that make up the galaxy, it is decidedly ordinary.

1 *The Pistol star, one of the largest and brightest stars in the Universe, is at the centre of this picture. Our Sun is around the same size as the many comparatively small stars surrounding the Pistol star.*

1

There are some exceptional stars littered across the cosmos: giant stars that shine 1,000 times as brightly as the Sun; neutron stars, which are so dense that they weigh the same as our Sun but are only 10 kilometres across; magnetars, whose magnetic field is so strong that, from close up, they would instantly kill you by rearranging every atom in your body. There are stars that pulsate, stars that explode, stars that glow with the merest dim, reddish light. But our Sun, at 4.6 billion years old, is stuck in a stable and unspectacular middle-aged rut. Even its official classification – a yellow G2 dwarf – sounds pretty boring.

If, for example, our Sun were the size of Betelgeuse, the giant red star in Orion, its surface would extend all the way out to Mars, completely engulfing the Earth. But such gigantic stars are extremely rare; the vast majority of stars have a diameter of around a million kilometres, roughly the size of the Sun. And although some stars are over 100 times more massive than the Sun, and others sustain themselves on just one-fifth of its mass, our own star's mass is also very close to the galactic average.

In fact, whatever comparison you want to choose – brightness, size, temperature, even age – ours is a disappointingly average star. But we shouldn't complain too much. If the Sun had been a very bright star, it would have burnt its fuel so fast that it would have lasted for only a million years...and we would not be here at all.

The Sun's nuclear
engine revealed

In 1905 Albert Einstein unveiled his revolutionary theory of special relativity, which described how strange things happen when you try to travel close to the speed of light. Almost as an afterthought, he scribbled an equation: $E = mc^2$ (the **e**nergy held in matter is equal to its **m**ass times the speed of light, **c**, squared). Aside from being the only mathematical equation that can be said to be truly famous, it introduced the world to a fantastic new idea: locked within matter was a huge store of energy. Even the smallest grain of sand holds more energy than the blast of many tonnes of high explosive, if only there were a way to release it.

Soon afterwards, studies of radioactive elements began to show that there might be a way to get at that energy. When an unstable element such as uranium decays, it splits into two smaller elements. But early researchers noticed that the sum of the parts is less than the whole – some of the mass of the original uranium atom is lost, being released as energy. In 1926 a British astronomer called Arthur Stanley Eddington was the first to see the solar potential in the conversion of mass to energy. An almost limitless reservoir of energy was locked inside the Sun's most abundant gas, hydrogen.

Eddington and his colleagues theorized that at the heart of the Sun, hydrogen was being converted to helium. The intense pressures and searing temperatures at the Sun's core were forcing hydrogen nuclei together, fusing them into helium. This nuclear transformation results in only a tiny loss of mass, but as Einstein had proved, even a tiny amount of matter contains an enormous amount of energy. Hard proof would come in the 1950s, when the devastating power of the hydrogen bomb was finally unleashed, but three decades earlier Eddington calculated that this still-hypothetical fusion of hydrogen could stoke a furnace that would rage for billions of years without showing any sign of shrinking. Even though 4 million tonnes of hydrogen are consumed by the Sun's nuclear furnace every second, it still takes 25 million years before the mass equivalent of our planet is used up – and that leaves another 300,000 Earth masses to go before all the hydrogen is exhausted. This raises an immediate question: if at the heart of the Sun there is a hydrogen bomb many times larger than all the planets put together, why does the Sun not explode? The answer is that the sheer weight of gas surrounding the core keeps the lid on this nuclear cauldron and prevents it from blowing itself up.

From the 1920s to the 1950s a collection of brilliant nuclear physicists worked out the details of Eddington's nuclear fusion reaction that powers the Sun. They theorized their way to the heart of our star and found, as he had predicted, that at the centre of the Sun lies a nuclear generator approximately 400,000 kilometres across, taking up about a quarter of the Sun's diameter. In this core, at temperatures of 15 million degrees Centigrade and pressures 225 billion times what we experience here on Earth, hydrogen is turned into helium, unleashing a steady stream of energy. These predictions were turning out to be extremely accurate. But while nuclear theorists refined their calculations, other astronomers were discovering that the internal structures of our Sun are anything but simple.

BELOW *Albert Einstein (1879–1955), physicist, enjoying some thinking time.*

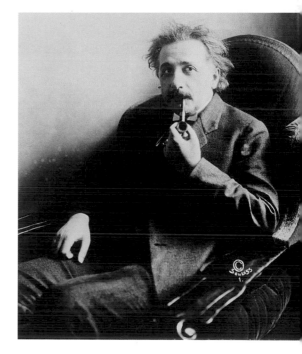

Fusion – the elusive power of the Sun

1

At the heart of the Sun lies the most efficient generator known to science. It is a nuclear reactor that fuses together the atoms of the most abundant element in the Universe, hydrogen, to form the second most abundant element, helium. During this transformation, tiny quantities of hydrogen release so much energy that if scientists could recreate a miniature version of the Sun's nuclear core here on Earth, it would solve all our energy needs at a stroke.

Of course, we have been generating nuclear energy for several decades, but our comparatively feeble reactors work not by fusing atoms but by splitting them: nuclear fission. In comparison to the highly productive and clean reactions taking place within the Sun, nuclear fission is a woefully inefficient means of producing energy and generates unwanted radioactive waste to boot. Nuclear fusion has been achieved on Earth, most notably in the hydrogen bomb, but the devastation caused by this terrible weapon only serves to highlight how dangerous it is to recreate the pressures and temperatures necessary to force hydrogen atoms together. Small tests in controlled fusion have been achieved under laboratory conditions, but the cost of safely generating energy in this way has, so far, outweighed the benefits.

Despite the difficulties, fusion is a dream worth pursuing. Imagine a world where virtually limitless energy can be released from a harmless, odourless gas locked in a chemical as freely available as water. This is the dream being chased by laboratories across the globe. One day they may succeed and the furnace that lies at the heart of stars will power our world.

1 The Sun's core on Earth. 'Ivy Mike' was the code name for this first ever detonation of a hydrogen bomb on 31 October 1952.

Windows on a tortured soul

Not only did Galileo move Aristotle's Sun from one part of the heavens to another, he also showed that it was far from perfect. With the advantage of the first astronomical telescope, Galileo found that the surface of the Sun was covered with dark blemishes (although he had no idea at the time that the Chinese had seen them as early as 28 BC). While Galileo was making his observations, a Jesuit priest from Germany was pursuing a long-term study of the

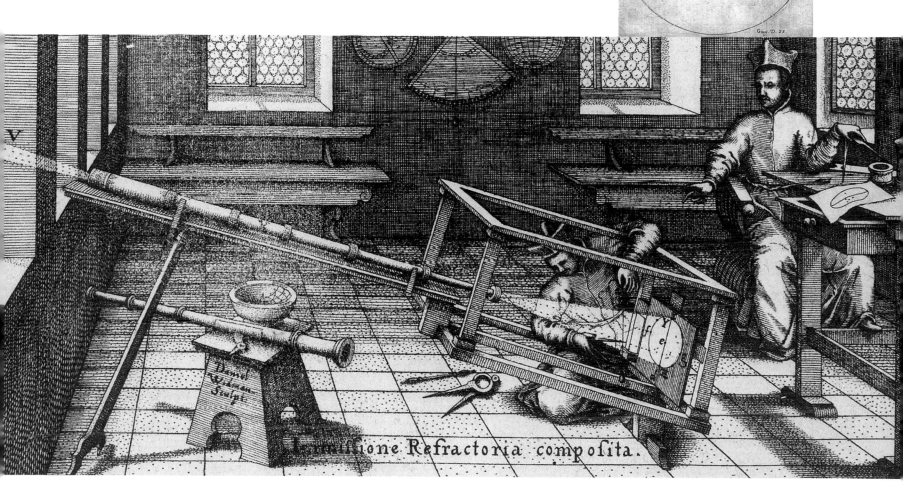

peculiar dark patches on the Sun. For over a decade, Christoph Scheiner watched the spots come and go across its surface. But the Jesuit, ever mindful of upsetting his fellow priests, not to mention Aristotle's legacy, explained the spots as tiny planets that orbited close to the Sun's surface. Galileo disagreed and fixed the dark patches firmly on the Sun. Galileo, as it transpired, was right. Sunspots are indeed part of the solar make-up.

In 1781 another famous astronomer, William Herschel, began a study of the Sun and was immediately intrigued by the occasional dark patches that Galileo had witnessed crossing

TOP *Illustration of sunspots from Galileo's 1613 book on the Sun.*

ABOVE *Christoph Scheiner (1575–1650) demonstrating the only safe method of viewing the Sun. The German Jesuit had a running battle with Galileo over the nature of sunspots.*

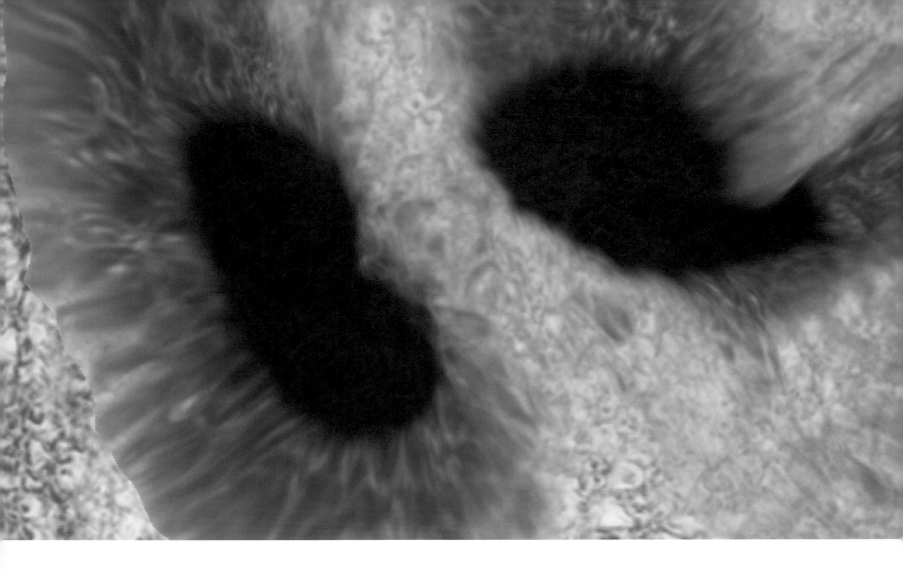

its disc. To Herschel the explanation was simple: beneath the Sun's searing cloud tops lay a cool planet. Sunspots were simply breaks in the cloud through which solid ground could be seen. In one respect Herschel turned out to be right — sunspots are indeed cooler than the surrounding chromosphere — but his notion of a solid planet was completely off the map. It's somehow reassuring to know that even the methodical and pedantic discoverer of Uranus was prone to the occasional romantic fantasy. Nevertheless, the spots had a story to tell and, half a century later, astronomers started to listen.

In 1829 a German pharmacist and amateur stargazer sold his business to become a full-time astronomer. Heinrich Samuel Schwabe's mission in life was to discover the planet Vulcan, a mythical planet (as it turned out) believed to lurk within Mercury's orbit. Schwabe's immediate problem was to avoid confusion between the transit of the dark spot of Vulcan and the passage of sunspots. For 17 years he meticulously plotted sunspots and in that time noticed a pattern to the appearance and frequency of these blemishes. In 1843 he announced that the numbers of sunspots seemed to grow over an 11-year period. Then they would suddenly disappear, only to start the process all over again. Sunspots had yielded one of the first hints about the incredibly complex behaviour of the Sun: our star is governed by an 11-year cycle.

In the mid-19th century another amateur astronomer of independent means immersed himself in his solar hobby-horse. Richard Carrington used a sizeable chunk of his

father's money to build a telescope in Redhill, England, and made an important contribution to our understanding of the Sun. Over the course of the 11-year cycle, Carrington noticed a hitherto unrecorded aspect of sunspot activity: the spots nearest the equator moved faster than the spots nearer the poles. Galileo had discovered that the Sun spun once on its axis every 27 days (when you take into account the rotation of the Earth, the actual rate is 25.4 days), but Carrington's discovery made it clear that the Sun didn't rotate as a single solid body. Like the giant gas planet Jupiter, the bands at the Sun's middle move faster than the rest. But the real nature of sunspots was still a mystery, and the clues they had offered on the inner workings of the Sun were far from being understood.

Magnetic sun

By the start of the 20th century, George Ellery Hale had established three new solar observatories, and it was from his Mount Wilson Observatory near Los Angeles that he turned his attention to the study of the mysterious sunspots. He made several important discoveries. First, he was able to calculate that sunspots, though still a scorching 4,000 degrees Centigrade, were about 2,000 degrees cooler than the surrounding surface, thus confirming the suspicions of earlier astronomers. He also saw that spectroscopic lines emanating from these sunspots bore the tell-tale signature of being distorted by a strong magnetic field. Hale confirmed that the Sun was not merely magnetic, but incredibly so; at some points the magnetic field was around 6,000 times stronger than the field that surrounds the Earth.

Later studies showed that pairs of sunspots – most sunspots form in pairs – were of opposing magnetic polarity. (Sunspots act in the same way as magnets, which also have a north and a south pole. Opposite poles attract and stick together, and 'like' poles – north with north, or south with south – repel each other.) Intriguingly, the sunspots leading the advance across the surface in the northern hemisphere of the Sun were of opposing polarity to those in the south. In 1913, as the Sun was about to start a new cycle, Hale was stunned to discover that the spots in both hemispheres had swapped polarity: the Sun's huge magnetic field had flipped!

Unlike the magnetic field on Earth, where the field lines run tidily from one pole to the other, the Sun's become unruly at the height of its 11-year cycle. Eventually, an English astronomer, Horace Babcock, devised a process that might explain this bizarre behaviour. At the beginning of the Sun's 11-year cycle its magnetic field lines run directly from, say, north to south. But because the electrically charged gases at the Sun's equator flow faster than at the poles the magnetic field crossing the Sun's middle is pulled out of shape. As the flows of gas within the Sun deform the central field lines, they twist around each other to form a braid. As the Sun turns, these braided lines lap their original position, while their feet stay firmly rooted to the poles. As the cycle progresses, the lines are wrapped around and around the Sun's equator and soon become even more intertwined. Eventually, some of these contorting field lines break out on to the solar surface. At the points at which they break through and at which they dive back

BELOW *Lines of chaos: over the course of its 11-year cycle, the Sun's magnetic field gets twisted into a furious jumble responsible for sunspots, prominences, flares and coronal mass ejections.*

down into the Sun, the temperature drops and the photosphere darkens. When this happens, we see a single pair of sunspots. After 11 years, the magnetic lines are so stretched and contorted that they suddenly snap and reorganize themselves back into neatly ordered lines, but this time running from south to north.

We now know that the Sun is smothered in hundreds of magnetic field lines which crisscross the surface like a haphazard jumble of overlapping croquet hoops. It appears that these looping magnetic arches are the lattice along which fiery prominences are channelled – they are the mysterious force guiding the tongues of flame back into the solar surface.

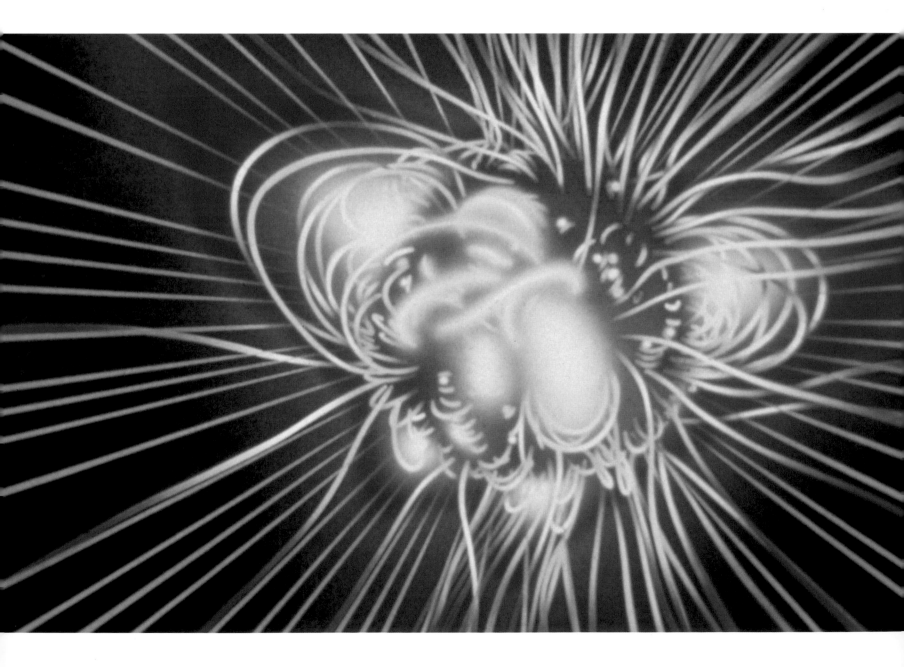

Mystical lights

Not all the prominences witnessed by Victorian astronomers were pulled back into the Sun. Occasionally, one of these tongues of fire would break away, dragging a huge ball of plasma (a cloud of superheated ionized gas) free of the Sun's influence to hurl itself out into the Solar System. These solar flares were first spotted by Carrington in 1859 and, once again, sunspots seemed to play a key role. Later, Hale witnessed several massive eruptions occurring in the vicinity of an angry group of sunspots. But Carrington and Hale were becoming aware that something else was happening whenever sunspots and prominences stormed across the Sun – something far removed from the Sun's surface. When solar activity was at its height, strange lights danced across the skies at the Earth's poles. At first it seemed too fantastic to contemplate, but somehow the Sun seemed to reach out and touch the Earth.

For thousands of years, people living near the Earth's poles have gazed in wonder at the aurora borealis at the North Pole and the aurora australis at the South Pole. Seen on clear nights, the auroras appear as dazzling sheets of multicoloured lights that seem to ebb and flow across the sky. The Danes kept records of auroral activity in Greenland for centuries, but no one really knew what was causing the displays. With the dawn of the space age, the answers were about to come thick and fast.

On 10 October 1946 the first scientific instrument was fired into the upper atmosphere using a captured German *V2* missile. It rose 90 kilometres above the Earth – not quite beyond our atmosphere, but near enough. The rocket carried equipment that sensed our planet was being bathed in ultraviolet light. Before the 1950s were out, other rockets discovered that space was alive with X-rays and gamma rays all emanating from the most active regions of the solar surface. Space was far from being a void; it was a hostile place teeming with deadly rays. Why could none of this radiation be detected from the ground? For the first time scientists had an inkling that the Earth's atmosphere was in some way protecting us from the worst of the Sun's excesses. But they were soon to discover that there was more to the Earth's defences than just the atmosphere.

ABOVE *Seen from above its pole, this image shows how the Sun's magnetic field lines get twisted by the faster rotation of its equator.*

Solar wind

Astronomers had long suspected that the Sun's corona might extend further than the 3-million-kilometre halo that had been seen during eclipses. Way back in the 7th century the Chinese had noticed that comets' tails don't lie directly in their wake but always point away from the Sun, regardless of the comets' direction. For a while it was thought that this effect was caused by some form of solar radiation, but by the late 1950s, rockets had ventured far enough into space to measure the Sun's radiation. Deadly it may be, but it isn't muscular enough to bend the tails of distant comets. Something else had to be out there. Several astronomers theorized about the possibility of high-speed particles streaming from the Sun, but none of the sums worked out. Then, one day in 1958, a young researcher at the University of Chicago sat down at his desk and a few hours later had made the numbers add up.

Eugene Parker postulated that the Sun was evaporating into space, producing a powerful flow of particles streaming constantly outwards: he called this 'solar wind'. Parker tried to publish his paper in the *Astrophysical Journal,* but the idea seemed so preposterous to the establishment that it was dismissed out of hand. The young astrophysicist drew fire from all sides and academic referees said that his paper was nonsense. The prevailing wisdom of the day was that the corona was a static feature, an extremely rarefied atmosphere that hugged the Sun in the same way that our atmosphere clings to the Earth. In Parker's model, at around 2 million kilometres from the Sun – the very point where the corona seemed to disappear – particles were accelerating away at 650,000 kilometres per hour. By the time they had reached the Earth, Parker estimated, they had reached a speed of around 1 million kilometres per hour. He envisioned the Sun's atmosphere continuing way out beyond the Earth and Mars, perhaps stretching even as far as Jupiter.

Parker did not have long to wait before the space age proved his opponents wrong. The first evidence was indirect, but nevertheless crucial. In the same year that Parker proposed his idea, the Americans had joined the space race in earnest. Whereas the Soviet satellite *Sputnik* was little more than an orbiting radio, the first US satellite, *Explorer 1*, carried genuine scientific experiments into space and they immediately hit paydirt. *Explorer 1* carried instruments designed by a scientist called James Van Allen to measure the cosmic rays flowing in space. At the time, scientists had expected to find interplanetary space riddled with particles from other stars and distant galaxies, but our Sun was not expected to contribute anything. What Van Allen found was quite unexpected. High above the Earth's atmosphere his instruments discovered that the magnetic field of the Earth spread far out into space. This magnetic envelope became known as the magnetosphere. Other on-board detectors registered

BELOW *The spectacular aurora borealis, or northern lights, seen here from Spitsbergen, a remote island in the Arctic Ocean.*

layer upon layer of charged particles that seemed to flow along these newly discovered magnetic field lines. These charged layers were named the Van Allen Belts. But where was the source of this dense sea of particles? Suddenly Parker's solar wind looked more promising.

On 12 September 1959 the Soviets launched *Lunik 2*. The probe was destined to crash-land on the Moon, but as it left the Earth behind, the particle detectors unexpectedly sprang into life. Something was out there: streams of charged particles flowing from the Sun at a speed of at least 60 kilometres per second. It seemed to confirm what Parker had predicted, but the Soviet measurements were short-lived and detractors of solar wind were still not convinced. The clinching evidence came in 1962 when, en route to Venus, the American probe *Mariner 2* proved once and for all that the fluctuating flow of charged particles measured earlier were, in fact, Parker's solar wind. Later measurements revealed that this stream of highly energetic particles – hydrogen, helium, protons, electrons – pour from its polar regions and spiral outwards in all directions as the Sun spins, looking like the spray from a rotating garden sprinkler. The Sun's 'atmosphere' extends way beyond the confines of the visible corona; in fact, you could say that the corona is just the visible part of the solar wind.

A port in a storm

The space age had barely begun and already we had learnt what an unpleasant place space really is. The vacuum that we had expected to find was far from empty. It was filled with a constant barrage of deadly gamma rays, X-rays and ultraviolet light bursting from the Sun at the speed of light; then there were the comparatively snail-like gusts of charged particles that billowed between the planets. What effect was this vicious combination having on the worlds that lay in their path? Eyes turned to Mercury, the planet that bore the full brunt of our Sun's deadly cocktail. At such proximity to the Sun, the endless barrage of energetic particles directed at Mercury make life impossible. How is it, then, that we on Earth are shielded from this deadly gale?

After *Explorer 1* had discovered Van Allen's belts, a stream of Russian and American satellites pieced together the various parts of an eternal drama played out high above our heads. At the upper reaches of our atmosphere a war rages between the Earth's defences and the deadly outbursts from the Sun. Streaming outwards from the poles, the Earth's magnetic field lines loop through space and rejoin our planet at the opposite pole. Our apple-shaped magnetosphere is nothing less than a force shield that repels the solar wind and channels it harmlessly around and away from our planet. Sometimes the solar particles find holes in the outer magnetosphere, only to be deflected by inner magnetic field lines and coralled towards the poles. These are the belts of charged particles that Van Allen discovered. At the poles – the only places where our field lines touch the Earth – waves of these particles flood into our atmosphere and create the dazzling auroras. Occasionally, a gust of particles manages to breach the Earth's force shield and we can suffer anything from the collapse of communication networks to the failure of power grids causing blackouts in major cities. But the Earth's magnetic shield does far more than protect our mobile phones: we need only

OPPOSITE *Composite image of the Sun taken in ultraviolet, showing its outer atmosphere or corona and solar wind. The black band shows the edge of a filter.*

ABOVE Apollo 12 *astronauts Al Bean and Pete Conrad carried out a series of experiments to determine the effects of solar wind on the Moon.*

glance nervously at our neighbours Mars and Venus to see that the shield is essential to our very existence.

When *Mariner 2* went to Venus in 1962, scientists were surprised to find that the Earth's twin planet has no magnetic field. This is almost certainly due to the planet's incredibly slow rotation. Without a magnetic shield, particles from the Sun flow unhindered through its upper atmosphere and blow it away. Luckily, the atmosphere on Venus is thick enough to handle this, at least for now, but the Sun does blow a wispy trail of its atmosphere all the way out to Earth's orbit. Mars has only weak and highly localized magnetic fields – nothing like the global fields that hover above the Earth – and scientists now believe that roughly one-fifth of Mars' atmosphere may have been lost in the relentless 4.5 billion-year battle with solar wind. The realization that the magnetosphere protects our atmosphere makes it doubly important. Remember that our upper atmosphere protects us from solar radiation: ultraviolet light, X-rays and gamma rays. Without this layer the Earth would be permanently bathed in a mutating carcinogenic light. Astronauts, even in the comparative safety of low Earth orbits partially protected by the outer layers of our magnetosphere, have reported flashing lights inside their eyes caused by charged particles burrowing through the walls of their craft, their eyelids and on to their retinas.

That *Apollo* astronauts on the Moon – which has no atmosphere to protect it from the ravages of the Sun – survived this extreme environment was more a matter of luck than judgement. We now know that between the visits of *Apollos 16* and *17* in 1972, the Sun was particularly active and highly dangerous. Had the astronauts been out there during these multiple flares, it is unlikely that they would have survived the trip.

As the space age found its feet and probes travelled further afield, it became clear that the solar wind was seemingly endless. At Jupiter, *Pioneers 10* and *11* discovered that the solar wind reacts strongly with the planet's gigantic magnetic field, creating bands of charged particles up to a million times more dense than those in our own Van Allen Belts. In recent years, the Hubble Space Telescope has taken the most beautiful images of auroras on Jupiter, showing a deep blue ring of light circling the planet's poles.

At Uranus, nearly 3 billion kilometres from the Sun, the solar wind drags the rolling Uranian magnetosphere into a candytwist tail that extends for a further 6 million kilometres. Now, way beyond the orbits of Neptune and Pluto, the *Voyager* probes continue to register solar wind. Thanks to the *Voyagers*, we now know that the planets are not just bound to the Sun by its gravitational pull: they all dwell within the solar atmosphere.

BELOW *The Hubble Space Telescope took these pictures of auroras at the poles of Jupiter and Saturn.*

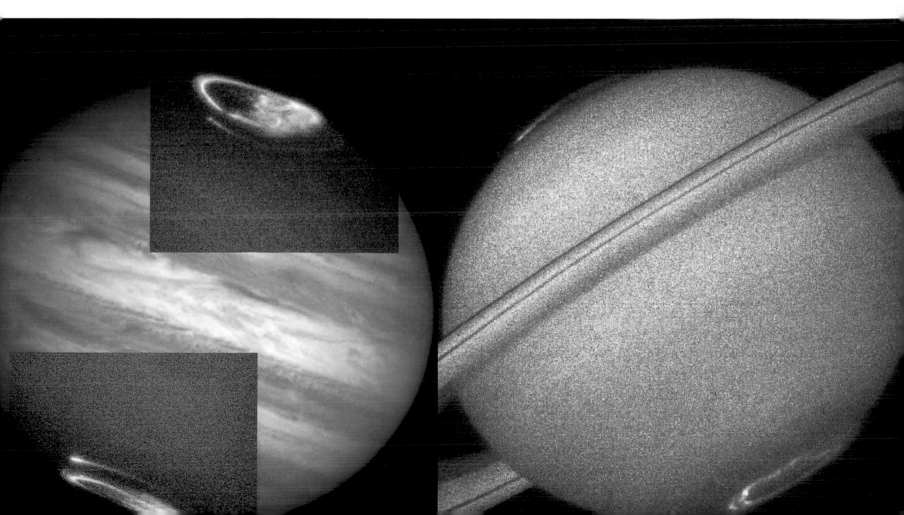

Angry Sun

In May 1973 NASA's *Skylab* was launched, and with it dawned a new era in solar science. *Skylab* was a manned solar observatory, a 75-tonne space-borne laboratory the size of a railway carriage, fashioned out of the final stage of a *Saturn V* rocket. For a total of nearly six months three three-man crews carried out 24-hour surveillance of the Sun. Only 40 per cent of the Sun's energy is emitted in the form of visible sunlight – the rest is in X-rays, gamma rays, ultraviolet light and infra-red light. Soaring 430 kilometres above the highest clouds, 1.5 million kilometres from the Sun, and sporting five X-ray and ultraviolet telescopes, two TV cameras and a telescope designed to create permanent eclipses by blocking out the solar disc, *Skylab* at last could study the Sun in its entirety. What Hale, Janssen and Lockyer would have given to be on board!

Before it had even arrived, the first crew was left in no doubt about the ferocious nature of the star it would be scrutinizing. During the unmanned launch of *Skylab*, the shield designed to protect the station from solar radiation was torn away. Drifting unprotected and in the full glare of the Sun, the temperature on board surged to 50 degrees Centigrade. A temporary shield had to be constructed and taken up with the first crew. Despite a successful deployment, the crew still had to endure several days working in temperatures in excess of 30 degrees Centigrade.

One day, early in their mission, the same astronauts witnessed an event that made their blood run cold. Over the course of a couple of hours a massive prominence bloomed out from the solar corona. The crew had already become used to gigantic loops of flame licking out from the surface, but this one was of a different order of magnitude altogether. Expanding at the rate of 2.5 million kilometres every hour, the ball of plasma soon seemed as broad as the Sun itself. Eventually, and without warning, the swollen mass exploded, spraying tens of billions of tonnes of solar plasma out into space. It looked like the Sun had fatally haemorrhaged, but as the scorching gases flew outwards into the Solar System, the Sun recovered and carried on its business as usual. The *Skylab* crew had become the first people ever to witness a Coronal Mass Ejection (CME); between them, the three crews would watch 100 more before their nine-month vigil was over.

Since the eclipse of 1839, the collective observations of the solar corona by Secchi, Carrington, Hale and their peers totalled just 80 hours. *Skylab* monitored the corona around the clock for almost nine months. In that time it amassed an incredible library of photographic evidence for the origins of the solar wind. By the time of *Skylab*'s launch, it had been pretty much established that particularly strong gusts of solar wind occasionally escaped through holes in the Sun's corona. The holes, in turn, were associated with gaps in the Sun's otherwise dense jumble of magnetic field lines. *Skylab*'s X-ray photographs of the corona clearly showed large black holes in the otherwise tightly bound layer. The holes rotate near the poles of the Sun. Ground-based studies of the Earth's magnetosphere showed that there was a typical lag of four and a half days between *Skylab*'s reports of these holes and the Earth being sprayed with charged particles. In addition, *Skylab* revealed that the larger the hole, the stronger the solar wind. The deadly nature of our Sun was clearer than ever before. From life-threatening

← Size of Earth

radiation to deadly solar wind, from solar flares to their terrifying big brothers, CMEs, this star in our midst needs to be closely watched.

   Skylab was abandoned early in 1974. More than three years later, a particularly ferocious sequence of solar flares heated the Earth's upper atmosphere, causing it to expand into the laboratory's orbit. The increased drag slowed the craft down and it began an inexorable tumble towards the Earth. *Skylab* eventually broke up somewhere over the Indian Ocean and showered debris across the Australian outback. In its brief tour of duty, *Skylab* served us well, not only increasing our understanding of the solar wind, flares and CMEs, but serving even in its death to remind us of the Sun's direct influence on our atmosphere.

No place to hide

Since *Skylab,* an international flotilla of unmanned solar observatories has been launched into space, making the Sun the most closely scrutinized body in our Solar System. At the time of writing, no fewer than seven spacecraft are monitoring the Sun and its effects on the Earth.

   The Earth, like all the planets apart from Pluto, lies in the same plane as the solar equator, so trying to understand the Sun from this narrow vantage point is rather like attempting to explain the behaviour of the Earth's atmosphere by studying the weather only at our Equator. In 1990 a joint European Space Agency (ESA) and NASA probe called *Ulysses* was sent on a long journey to the Sun's poles. Flying out to Jupiter, the probe used the gas giant as a gravitational slingshot to hurl itself out of the equatorial plane. First it swooped under the Sun's south pole, then it swung up around the other side of the Sun and over its north pole. This incredible journey gave the probe a unique vantage point, allowing it to build a picture of the solar wind from above and below. Before *Ulysses*, scientists thought that the shape made by the wind leaving the Sun was like a simple rotating garden sprinkler, but the probe has discovered that there are two types of solar wind: the fast wind that emanates from the poles, and a slower wind that peels away from the equator. These different winds streaming out of the

ABOVE *A huge solar eruption captured by* Skylab. *Compared to the Sun our planet is but a speck. Some stars are thousands of times bigger even than our Sun.*

OPPOSITE TOP *The first* Skylab *crew lifts off from the Kennedy Space Center, Florida, 25 May 1973.*

OPPOSITE BOTTOM *Inside the orbital workstation with the final* Skylab *crew.*

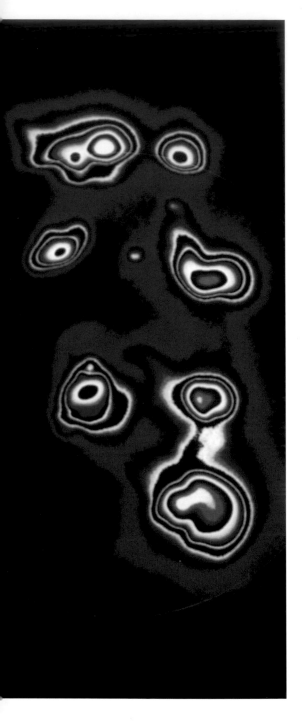

Sun push and pull each other in such a way that their combined forces create a complex undulating spiral pattern.

On 2 December 1995 ESA and NASA joined forces again to place a 2-tonne solar laboratory in space. Hovering at the point where the Sun's gravitational pull is exactly matched by that of the Earth, the *Solar and Heliospheric Observatory* (*SOHO*), 1.5 million kilometres from home, has been on 24-hour duty ever since. Armed with 12 precision instruments, this solar sentinel sends thousands of images of the solar surface every day to a small army of scientists based around the world. It is, by some margin, the most sophisticated instrument ever built to study our star.

SOHO's other job is to look deep into the heart of our star or, more accurately, to listen. Just as seismometers on Earth record the rumblings from deep inside our planet, so *SOHO* has been listening to the sounds bellowing from the Sun. Within its gaseous globe, the Sun rings, throbs and echoes with a chorus of 10 million separate voices. As the sound waves travel inwards, outwards, east and west, they cause the gases within the Sun, from the core all the way up to the surface, to vibrate. Instruments on *SOHO* monitoring the shimmering surface can pick out one by one the vibrations ringing, humming and pulsating inside the Sun, and has discovered some remarkable new secrets lurking deep beneath the surface.

Journey to the centre of the Sun

Fifteen thousand kilometres inside the Sun, *SOHO* has detected winds drifting from the equator to the poles. These winds move so slowly that they take an entire Earth year to finish their journey. Around the poles, just below the visible surface, *SOHO* has found vast rivers of plasma – a superheated gas – circling the Sun, much as our own jet streams circle the Earth. Images of the Sun in ultraviolet have shown it to sparkle like a multi-faceted ballroom globe. *SOHO* has seen white glowing pockets of plasma collect around the equator before CMEs burst out. As the twisting magnetic fields routinely slip and re-align themselves, these balls of plasma break free, like bubbles from the side of a tapped glass, and flow outwards to become the solar wind. *SOHO* has seen tornadoes of fire whipped up amid the mayhem that rages throughout the Sun's torrid atmosphere and recorded evidence for sunquakes – surface disturbances that ripple outwards from the centre of erupting solar flares.

But of all its discoveries, perhaps the most important has been *SOHO*'s unveiling of the Sun's hidden magnetic generator. Looking further and further into the Sun, *SOHO* revealed that the radiation zone – the layer that surrounds the nuclear core – rotates uniformly. Just above the radiation zone, however, the convection zone rotates differentially, i.e. faster at the equator than at the poles. Most scientists believe that it is here, at the junction of the radiation zone and the convection zone above it, that the shearing forces create massive electrical currents, which in turn give rise to the Sun's powerful magnetic field.

Tantalizingly, *SOHO* has managed to see its way down to the very edge of the Sun's nuclear core. The hope was that the more it told us about the core, the closer we would get to understanding its magnetic clock and why it has an 11-year cycle; for a while, it seemed like it would never get the chance. At around midnight on 25 June 1998, during routine maintenance operations, ground controllers lost contact with *SOHO*. It has since been recovered and slowly coaxed back into life. Not only did *SOHO* survive its brief period of hibernation, but some of its instruments are working better than ever. *SOHO* is once again gearing up for the next peak in the solar cycle. In the year 2000 the Sun will be bursting at its seams, hurling matter throughout the Solar System and bubbling frantically for several years before its temper cools off and the planets are offered another 11-year respite. What insights *SOHO* will offer hungry solar scientists into the machinations of our star at the height of its activity can only be imagined.

The stretching boundary

How far does the Sun's influence extend? In 1977, when the two *Voyager* craft set off on their Grand Tour of the gas giants (see Chapter 4), one of their objectives was to measure the solar wind. At that time the solar wind was generally expected to peter out at the orbit of Jupiter. Now, over two decades later, the *Voyagers* are still going strong and still detecting solar wind. Hurtling directly away from the Sun in different directions at 40,000 kilometres per hour, it is estimated that by 2015 they will reach the heliopause. This is the invisible line in space at which the Sun's weakening wind runs out of steam and is held in check by the faint whisper of interstellar wind that fills the space between the stars. It is the very limit of our Sun's influence – the frontier of its empire. In 1998, massive eruptions from the Sun caused interplanetary mayhem. It was the most explosive display of solar activity ever witnessed and some months later, though separated by more than half a billion kilometres, both *Voyagers* reported being overtaken by strong solar gales. Over a year passed before the *Voyagers* sent back startling news. They had just encountered, in the form of radio waves, the ghostly echo of the two massive solar gusts passing in the other direction. The wind had raced on to the heliopause, where it met the outer envelope of the Sun's magnetic field. As the charged particles ploughed into the magnetic field, they generated radio waves, which were picked up by the *Voyagers*. From this evidence, scientists now estimate that the point where the solar wind is constrained by the equal forces of interstellar wind is roughly 22 billion kilometres from the Sun – 15,000 times further away than the Earth. The messages from the receding *Voyagers* are already feeble and no one can be sure that when they eventually cross the edge of the Solar System, the report they send back will be heard. In all likelihood, as they slip from our Solar System and into the eternity of space, they are destined to keep their secrets to themselves.

ABOVE *Artist's impression of* SOHO – *the greatest solar observatory in history.*

atmosphere

IT IS NOVEMBER 1971, AND SOVIET scientists are bracing themselves for an encounter with Mars at 25,000 kilometres per hour. *Mars 2* and *3* are the first robotic craft to attempt to penetrate the Martian atmosphere. Reports from telescopes across the country all bear the grim news of a giant storm on Mars. The planet is covered from pole to pole in a thick blanket of dust. But there is no reprogramming the probes — *Mars 2* is first to take the plunge and 7 kilometres up from the surface its data stream flinches from a violent buffeting. Moments later, the engineers watch the readings blink out, and they

can only imagine the metallic thud with which the probe must have hit the surface. Five days later, *Mars 3* also takes a beating on the way in. Then, suddenly, the altitude readings level out and a ripple of muted excitement spreads around the control room. Right now, on the surface of another planet, four sandblasted petals must be unfurling, an antenna should be searching the alien sky for its home planet, and a TV camera will be stirring into life. Ninety seconds later, there's a huddle of scientists around the teleprinter as the picture slowly begins to emerge. But then, after just 15 seconds, the transmission stops. That glorious first picture of the surface of Mars is nothing more than a fuzzy mess of squiggles. The probe is dead. Nobody can be certain of the cause, but the dust storm is the prime suspect. Mankind's first contact with the atmosphere of Mars is over, and the welcome was anything but warm.

It was 1957 when man first escaped the Earth's atmosphere. In that year, US Air Force pilot David Śimons reached 37,000 metres riding on *Man High Mission 2*. Viewing the vulnerable haze of gases from a pressurized capsule slung below an experimental balloon, Simons was the first human to see beyond the air that has contained us since the beginning. As an old man, he still recalls the vision as 'the most startling, the most arresting sight I have ever seen – sunset through the pristine clarity of the edge of space'.

PAGES 166–7 *Saturn's moon Titan, the only moon in the Solar System with an atmosphere.*

The edge of space

In truth, our atmosphere is a paltry, thin veneer of gas, barely worth mentioning in the measure of a planet such as the Earth. If it could be cooled enough to freeze solid, all the air we have would shrink down to a layer just 15 metres deep around our globe. From our humble human perspective, though, the atmosphere means much more to us than that. It is a giant storm, 4,500 trillion tonnes of air, swirling above our heads. It is responsible for a world of life-threatening extremes – freezing blizzards in Antarctica, desert sandstorms in the Sahara, hurricane winds in the tropics and flash floods across the flatlands.

ABOVE LEFT *Sunrise from the edge of space, photographed by space-shuttle astronauts 35 years after David Simons first witnessed this sight.*

ABOVE RIGHT *Twenty-mile high club: one of the first true space travellers, David Simons made the cover of* Life *magazine when he took this picture of himself dangling above the Earth's atmosphere in a helium balloon in 1957.*

But without our atmosphere we would be lost. By itself, the Sun's heat would not be enough to melt water, except in a narrow band around the equator. The greenhouse effect is not just a 20th-century phenomenon – it's been with us for billions of years and is first and foremost our saviour, adding 30 degrees Centigrade of warmth to our world, tipping the fine balance between frigid death and comfortable existence. But our atmosphere does more than sustain life: it provides the overarching experience of being alive. Our waking day is dominated by the sky that hangs above our heads, mostly clear, pellucid blue, sometimes a deep purple, for brief instants even the purest orange. It can be filled with fantastic and impossible structures, clouds high and wispy or low and ominously dark. But even if we had no eyes to see, the Earth's air would still dominate our existence. The breath in our lungs, the wind in our hair, the rain on our face, the feeling of heat or cold – every moment of our life is affected by an awareness of our atmosphere.

Little wonder, then, that as soon as it was possible to venture away from our planet towards the other worlds that orbit the Sun, it was their atmospheres, their weather, the promise of experiencing life below alien skies that fascinated us beyond all else. This is the story of our first breaths of extraterrestrial air.

A pierced veil

I have often wondered that when I have viewed Venus…she always appeared to me all over equally lucid, that I can't say I observed so much as one spot on her. Is not all that light reflected from an atmosphere surrounding Venus?
Christiaan Huygens, *New Conjectures Concerning the Planetary Worlds,* c.1690

From the earliest days of civilization, the planets were seen as wandering stars in the night sky, disembodied patches of moving light, nothing more. With the invention of the telescope in the early 17th century came the realization that the planets were solid bodies, tangible and firm, something like the Earth. But Dutch astronomer Christiaan Huygens, a master of optical instruments, mathematics and astronomy, was the first to consider them as worlds – places that might have air, oceans and even inhabitants. He gazed at Mars, Jupiter and Saturn, but the bright morning star of Venus was the place that most interested him: could there really be an atmosphere, a sky, on our neighbouring world?

It took the best part of a century for Huygens' conjecture about Venus to bear fruit. In 1761 Russian scientist Mikhail Lomonosov was using his telescope at the University of St Petersburg to watch Venus pass across the face of the Sun – a rare astronomical alignment called a 'transit'. As the tiny silhouette of Venus inched across the glaring disc of the Sun, he noticed that the edge of the planet wasn't hard and crisp like the Moon's was during an eclipse – Venus seemed fuzzy. Then, as the planet crept off the disc, he was stunned to see the curved edge of the planet, the limb, glowing like a halo. His deduction of 'an atmosphere equal to, if not greater than, that which envelops our earthly sphere' was spot on.

A world similar in size to the Earth, with an atmosphere, and only slightly closer to the Sun… So tantalizing was the idea of clouds on Venus that it wasn't long before Lomonosov's sighting sparked off an analogy with Earth. Everything about Venus suggested that it, too, was a life-cradling oasis. Over the course of the next two centuries, it would become in the mind of man a misty haven, a swampland teeming with exotic animals and plants – an Eden.

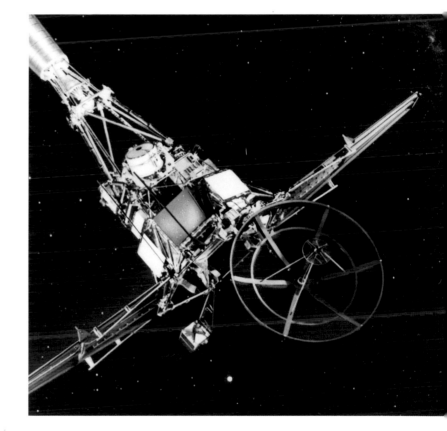

The first signs that this idyllic picture might be wrong came in the early 20th century. Astronomers used spectroscopes to dissect the chemical make-up of the atmosphere on Venus. They found no oxygen, no sign of anything in fact, other than large amounts of carbon dioxide. Then, in the 1950s, scientists using newly built radio telescopes trained their dishes on the morning star and discovered that Venus must be incredibly hot – hundreds of degrees Centigrade – so intense were the natural signals it was radiating. The readings were in such contrast to the accepted image of Venus that many couldn't believe them. Instead they imagined that the intense radio signals might be generated high up in the atmosphere by some unknown mechanism, and that safely below the clouds the surface might still be a balmy oasis. Then another idea emerged: all that carbon dioxide could be acting like an insulating blanket around the planet. Venus's surface might be sweltering due to a massive greenhouse effect, adding not 30 degrees, as on Earth, but more like 300 degrees Centigrade to the temperature bestowed on it by the Sun.

So it was, as the space age dawned, that Venus, not Mars, was the prime destination, the source of so many mysteries. In the winter of 1962 *Mariner 2*, the USA's first interplanetary probe, turned its detectors on this shrouded world from a comfortable distance of

ABOVE Mariner 2 *was the first successful interplanetary voyager. The craft failed and recovered so many times on the way to Venus that some quipped that JPL (the Jet Propulsion Laboratory in California running the mission) stood for 'Just Plain Lucky'.*

35,000 kilometres. *Mariner 2*'s readings were not totally conclusive, but everything seemed to point to the heat emanating from the surface rather than the cloud tops. And the temperature clocked in at an incredible 400 degrees Centigrade.

Then, on 18 October 1967, *Venera 4* plunged right into the cloud tops of Venus. As the Soviet entry probe pulled its parachute cord and began a leisurely descent towards the surface, it sent back the first details about conditions beneath the Venusian shroud. Soviet engineers had no idea how long the descent would take, so after 94 minutes, when the altimeter stopped and the craft fell silent, for a few glorious moments they thought that it had landed. Then, when they scrutinized the last few minutes of its transmission, they discovered the disheartening truth. At the instant it had seemed to land, the craft had still been 25 kilometres above the ground. In its last message home the temperature reading was hotter than an oven, but, more surprising still, it had been registering pressures of 18 atmospheres – 18 times greater than at sea level on Earth. At that moment its hull gave way and *Venera 4* was crushed into a pile of wrecked metal.

The following day, while *Venera 4* lay half-molten on the surface, a second American craft flew within 4,000 kilometres of Venus. *Mariner 5* confirmed the high temperatures beneath the clouds and estimated that the pressure at the surface might be a staggering 100 atmospheres. Both probes reported that the make-up of Venusian air was wholly unearthly – 95 per cent carbon dioxide. Venus was no sister of ours.

A tough nut to crack

In the spring of 1969 *Veneras 5* and *6* went the way of their predecessor: neither got closer than 16 kilometres above the surface before being crushed. It was clear to the Soviets that landing on Venus was going to be a supreme challenge of engineering. The task fell to the Lavotchkin Association in Moscow. This was the centre that had designed *Luna 9*, the first probe to soft-land on the Moon, and in the late 1960s it was already building the ambitious *Luna 15* 'moonscooper', the lunar sample-return spacecraft which attempted to trump *Apollo 11*. Lavotchkin's engineers were the heroes of the Soviet space programme and they were not about to be beaten by Venus.

When *Venera 4* was launched, the Lavotchkin team had guessed that the pressure on the surface of Venus might be at most ten times that at sea level on Earth. They weren't just wrong about that. They'd also been prepared to land in a Venusian ocean and had made sure the probe could float, adding an antenna that would separate itself from the probe by means of a dissolve-on-contact sugar-lock. With all that they now knew about Venus, their preparations looked like a joke. Given the temperatures, there was no chance of liquid water on Venus, and the pressures were many times higher than they'd predicted. Vladimir Perminov, the man chiefly responsible for building the new probes, ordered the construction of a special high-pressure chamber and a giant centrifuge to simulate the huge braking forces the probes would experience when they first slammed into the atmosphere at 11 kilometres per second.

Perminov had discussed with his colleagues what the likely pressure on Venus might be; most of them estimated it at between 60 and 100 atmospheres. Perminov decided to play it safe and design a spacecraft to withstand 180 atmospheres. Prototypes for *Venera 7* were soon undergoing their brutal tests. The pressure in the chamber was so great that it took an entire day of decompression just to get the door open. Venus wasn't going to get the better of Perminov's little probe.

Venera 7 arrived at Venus on 15 December 1970. Its parachutes had been trimmed for it to fall fast and minimize the amount of time spent in the cruel atmosphere before landing. For a while it appeared that all the testing had gone to waste. *Venera 7* transmitted for a mere 35 minutes before it fell silent. At first, mission controllers concluded that the probe had gone the way of its forebears. The radio signal screeched white noise, but suddenly one radio operator heard something else in the signal. Barely discernible above the sizzling silence, *Venera 7*'s muted call was still there and the pressure and temperature readings were stable. There could be only one conclusion: the craft had stopped falling and was transmitting from the surface. Perhaps it had fallen over as it landed, weakening the signal. No matter, it was there! For 23 minutes, a group of scientists huddled in ecstatic silence, listening to the first radio broadcast from the surface of another planet.

But if the news of their success was cause for celebration, the message from the surface was sobering. There was no more room for hope – the dream inspired by Lomonosov was over. Here is a planet that ought to be somewhat like home, but its atmosphere is made up almost entirely of carbon dioxide rather than containing oxygen and nitrogen like ours. The pressure on the surface is 90 atmospheres, equivalent to being a kilometre down in the Earth's oceans. The temperature, 470 degrees Centigrade, is hot enough to melt tin, lead, zinc…organic matter wouldn't stand a chance. How unlike the Earth was it possible to be?

ABOVE *Venera 7 had no cameras or soil scoops, but it could handle tremendous heat and pressure. In 1970 it became the first craft to survive the scorching descent through Venus' atmosphere and successfully land on its surface in working order.*

OPPOSITE *These images of Venus were taken through* Pioneer Venus Orbiter's *ultraviolet light – the only way to observe cloud detail on the planet.*

The other side of the tracks

Well before the era of space travel it was clear that the air on Mars couldn't be as thick as Venus' shroud. In 1656 Christiaan Huygens put his optics to good use again and glimpsed white polar caps on the red planet, surely a sign that some atmosphere existed there. But it was William Herschel who first deduced that Martian air could not be very substantial. Riding high on his discovery of Uranus two years earlier in 1781, Herschel was now astronomer to King George III. He focused on Mars just in time to see it passing directly in front of a star. Expecting to see the starlight twinkle as it made its way to Earth via a sliver of Martian atmosphere, he was surprised to see barely any disturbance at all – there could not be much air there. Over the next century, though, reports of thin, wispy cloud cover came from many different astronomers. There was definitely something covering Mars, but its true nature was proving hard to pin down.

Mars' global weather

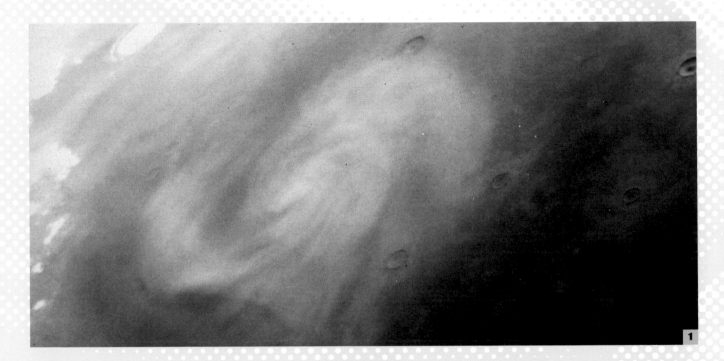

1

2

Although from day to day Mars can't muster anything like the variability of the weather systems on Earth, nothing on our world matches the scale of the global dust-storms that seem to coincide with summer in the southern hemisphere of Mars. When NASA's two *Viking* landers got to Mars, we finally had a chance to have weather stations on the surface, and scientists could begin to work out the causes of these giant squalls.

Because Mars spins on its axis with a tilt almost identical to the Earth's, it enjoys similar, albeit longer seasons. During the Martian summer, the midday temperature near the equator can soar to a momentarily comfortable 22 degrees Centigrade. Throughout the spring, as the temperature starts to rise, so does the equatorial air, leaving behind a zone of low pressure near the ground. This hole sucks in colder air from the polar caps. Sometimes the wind can start to blow at hundreds of kilometres per hour, and at those speeds even the thin Martian air has enough energy to pick up

substantial quantities of dust. Once these storms reach the equator, they are dragged high up into the atmosphere by the rising currents where the now reddened air makes its way back to the poles. Now the Martian weather goes into overdrive. As the dust particles in the upper atmosphere absorb more heat from the Sun, the increasing energy whips the winds still faster, dragging more dust up into the sky and sending the weather system into a vicious cycle. Occasionally, this scenario is so extreme that a blanket of dust swallows the entire planet. The winds starts to die down only when the dust in the upper atmosphere gets so thick that it stops the planet's surface from being warmed by the Sun.

1   *A giant cloud of dust swirls above the barren plains of Mars, captured by one of the Viking orbiters.*
2   *These images show the progression of a dust storm on Mars, photographed by the Lowell Observatory in Flagstaff, Arizona in 1973. The top image shows the planet just before the storm – a mere eight days later the entire planet was enshrouded in dust (bottom image).*

The first real answers came on 14 July 1965 when NASA's *Mariner 4* sailed within 10,000 kilometres of Mars. As described in Chapter 3, the mission hit the headlines with 21 black-and-white pictures of the cratered surface, but it also sent back news of the atmosphere. The initial estimates were that Martian air would be one-tenth the thickness of Earth's atmosphere, but when the results finally came in, the pressure on the surface turned out to be less than one-hundredth of an atmosphere. That meant the air was very thin, roughly the same pressure as the air outside David Simons' balloon at its highest point of ascent. Mars was a very different kind of planet from Venus, except in one striking regard. *Mariner 4*'s spectroscope showed large quantities of carbon dioxide — in fact, something like 90 per cent of Martian air was made of this gas. Suddenly, the Earth, with only a trace of carbon dioxide, 78 per cent nitrogen and 20 per cent oxygen, looked like the odd one out of the three.

Such a thin atmosphere might first have seemed like welcome relief compared with Venus, but to engineers trying to design a probe to make a soft landing it presented serious difficulties. The friction of first contact with the Martian atmosphere from the void of space would still demand a hefty heat shield, but any probe would have to take a long curving path downwards if the thin Martian air were to have any chance of slowing it down enough to make a safe landing. Nothing as simple as the ball-shaped *Venera* probes would do on Mars; a probe would need a finely engineered entry cone to keep it on the right path, and the final phase of the descent called for a giant parachute and retro-rockets. And then there were the huge, planet-wide dust-storms that astronomers had been watching from Earth for years. Any probe might have to contend with winds of hundreds of kilometres per hour.

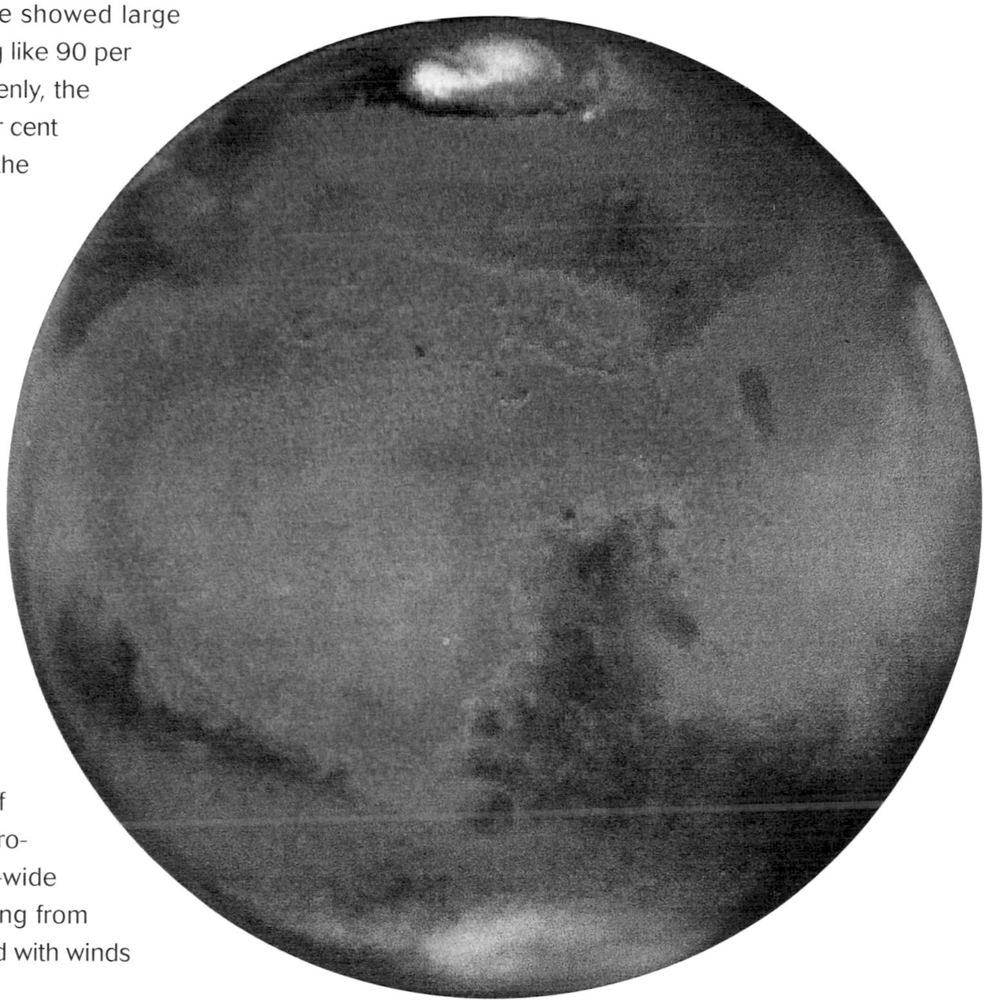

It was an engineering challenge the Russians attempted many years before the Americans. But, as we saw, in that winter of 1971 *Mars 2* and *3* were doomed to failure. A thousand and one things could have happened to those pioneering probes on the way down, but to this day mission scientist Mikhail Marov has no hesitation in blaming their demise on the global dust-storm. In his mind's eye he still sees the two probes hurled sideways by the raging winds, making a glancing approach to the surface and bouncing before coming to rest, dead or on their last legs. The sight of a Martian sky overhead remained an elusive vista.

ABOVE *Ice-caps on Mars, seen here in an image taken by the Hubble Space Telescope, tipped off early astronomers that Mars might have an atmosphere.*

In the summer of 1976 NASA was finally ready for an attempt to pierce the atmosphere of another world. The *Viking* Mars landing mission had entrusted the design of its heat shield to a safe pair of hands. Alvin Seiff was a veteran scientist at NASA's Ames Laboratory. It was at this research centre, just south of San Francisco, that he had designed the agency's first-ever heat shield. It had saved the likes of pioneering astronauts Al Shepard and John Glenn from a good roasting when they re-entered the Earth's atmosphere in their *Mercury* capsules back in the early 1960s. But Mars seemed a tricky proposition to Seiff. There were so many unknowns – the way the pressure changed with altitude, the exact chemical composition of the air, the amount of dust present, to name just a few – it seemed foolish to be attempting a soft landing before all those factors were known. Five years earlier, Seiff had proposed a mission to make a crash landing on Mars with a probe that would measure and radio back every detail about the atmosphere on its descent. But budget cuts cancelled that mission; now everything had to be perfect first time with *Viking*.

On 20 July 1976 *Viking 1* dropped out of orbit and barrelled into the upper atmosphere of Mars at 18,000 kilometres per hour. Within seconds the heat shield had reached 1,500 degrees Centigrade. Seiff's nose cone was doing more than just taking the heat; it was busy taking readings on the chemicals, temperatures and pressures he'd wanted so badly. When *Viking* was still 6 kilometres above the surface, a drogue chute fizzed out behind it. The main white canopy fanned out to slow the craft to a gentle 60 metres per second, until the retro-rockets lit up and eased it on to the dusty Martian surface.

Those first moments after landing were described in Chapter 3, but on day two the first colour pictures beamed back to NASA's Jet Propulsion Laboratory (JPL) in Pasadena, California, brought the sight that atmospheric scientists had waited for: the Martian sky. Before the mission, most of them had expected the sky on Mars to be basically black. To their minds the air was not thick enough to scatter the light of the Sun everywhere and create the characteristic blue light we see on Earth. Only at the horizon, looking through the Martian atmosphere on a slant, they reasoned, might there be enough air to cause a bluish tinge.

But when that historic colour panorama was released to the press, it showed a familiar scene: reddish soil overhung with classic Earthly blue. One of the *Viking* scientists, Jim Pollack,

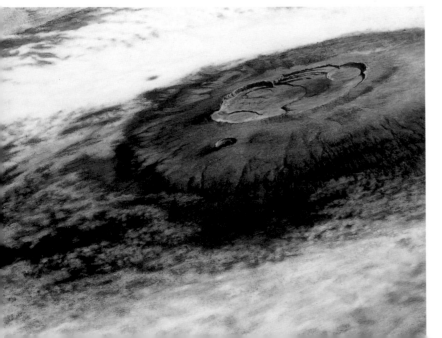

BELOW *Just as on Earth, moisture in the Martian atmosphere condenses out at altitude as warmer air rushes up the side of mountains, in this case the giant volcano Olympus Mons.*

was immediately suspicious and went back to check the raw data pumped out by the probe. Sure enough, he found that the image-processing team had, by instinct, artificially raised the intensity of the blues in the image so that the sky looked the way they'd always seen it – cool azure. Pollack ran off a new image, this time with the colours properly balanced. What he saw was another surprise: not the hazy black of space, but a sky of salmony-pink. Evidently, the Martian sky was coated with a permanent gauze of suspended red dust particles. Our first new sky was truly an alien sight.

And from this barren, rock-strewn plain that was now the permanent resting place of *Viking 1* came the first weather report from Mars. Winds were light and variable, around 20 kilometres per hour, which, due to the thinness of the air,

would feel to us more like a delicate waft of 2 kilometres per hour. The noontime high was minus 31 degrees Centigrade, the overnight low a frigid 100 below. There was no greenhouse effect on this world. The thin atmosphere was not sufficient to trap the raw heat of the Sun, which at this distance from our star meant that Mars was a frozen world. But the paucity of the air had one benefit: predictable weather, for the most part. Each day was much like the last, maybe slightly colder or warmer, depending on the season. *Viking*'s short-term forecast was: 'Fine and sunny with occasional scattered high cloud. Very cold and light winds. Further outlook similar.'

ABOVE *Blue or pink? The first colour picture of the surface of Mars sent back by* Viking 1 *was mistakenly processed back on Earth to show a blue sky (left). It wasn't until a few days later that the error was found, and the real picture (right) emerged.*

A   t a l e   o f   t h r e e   t e r r e s t r i a l s

Now that planetary scientists had had a taste of what Mars and Venus were really like, they struggled to understand how three rocky planets could have turned out so differently. Why is the atmosphere on Venus 100 times as dense as the Earth's, which is in turn 100 times denser than the air on Mars? How did Earth end up with a nitrogen/oxygen atmosphere instead of one consisting mostly of carbon dioxide? Solving these puzzles has taken many years of work by many researchers, but they now seem to have a story that explains the three planets' different fates. To tell it, we have to go back to the very beginning, to before the planets were even born.

As we saw in Chapter 1, during the early stages of planetary formation, all three embryonic worlds were probably very similar, their growing cores all grazing on the dust and asteroids swirling around our young Sun. But Earth and Venus, being further away from the disruptive tug of Jupiter, began to grow bigger than Mars. As the planets grew, their pull got stronger and the asteroids and lumps of dust rained down at higher and higher speeds until they were partially vaporized on impact, probably along with a fair proportion of the planets' surface. These impacting chunks were predominantly a mixture of rock and water ice. At first, the steam released on impact would drift off into space. But as the planets swelled in size, so did their gravitational pull and eventually the gas couldn't escape. Soon the nascent Earth and Venus were shrouded in a thick blanket of steam. At a temperature of around 1,500 degrees Centigrade, these were the first atmospheres of the inner Solar System.

By this stage, Mars had already gone down a different path. With only one-tenth the mass of Earth or Venus, Mars never managed to cause impacting asteroids to bombard it fast enough to be vaporized. Early Mars would have had a small amount of steam and some other gases in its thin air, and we know it must also have had water on its surface (see Chapter 7), but most of the volatile chemicals were still in the form of lumps of underground ice. As the red planet cooled, much of its early atmosphere ended up as a layer of frost trapped in the surface.

Mars was now stuck in a corner: the thinner the atmosphere is to start with, the harder it is to keep hold of. The early Solar System boasted a large family of asteroids left over after the planets finished forming, and during the following half billion years there were plenty of catastrophic collisions between the planets and these floating mountains. These devastating impacts – far larger than the one that is blamed for wiping out the dinosaurs on Earth – would blow huge sections of the planets' atmospheres into space. Mars, with its weak gravity and already thin atmosphere, would have suffered from this loss much more than Earth and Venus.

There is much debate about the exact sequence of events on Earth and Venus during those first half billion years, but the current best guess seems to be that as the planets cooled down, some of the thick veil of steam rained out on to the surface as reservoirs of water. This certainly happened on Earth. On Venus, closer to the Sun and always hotter than the Earth, if there ever were oceans, they would never have cooled much below boiling point. The atmosphere on Venus was continually saturated with steam. That was the critical difference that divided the destinies of these two planets.

Water vapour is the most efficient greenhouse gas known to science. It is transparent to visible light but reflects infrared light. As the Sun's rays fell on the young and steamy Venus, a small amount penetrated the fog and reached the surface, where it was absorbed. Hot rocks then radiated the Sun's energy back into the atmosphere in the form of infrared heat, which was completely sealed in by the thick veil of water. Venus stayed hot.

Meanwhile, high up in the atmosphere, ultraviolet light from the Sun was breaking water into its chemical constituents, hydrogen and oxygen. Hydrogen is so light that, once unshackled from the heavier atoms of oxygen, it simply floats off into space. Some of the oxygen would have been knocked out by the escaping hydrogen atoms, but most would join with carbon atoms in the hot surface rocks to form carbon dioxide, another potent greenhouse gas. Thus, although the oceans gradually boiled away into the sky and that shroud of steam slowly disappeared, the greenhouse effect never stopped. Steam in the atmosphere of Venus was slowly replaced with the massive amount of carbon dioxide that we see today. Venus is often referred to as a planet with a 'runaway' greenhouse effect, as if it had started off cool and then heated up. This is a bit misleading, since at first the planet was even hotter than it is now. But Venus was never able to get rid of a thick layer of greenhouse gas – either steam or carbon dioxide – from its skies. Venus didn't heat up; it just never had a chance to really cool down.

The Earth was just far enough from the Sun for the oceans to rain out, taking with them most of the greenhouse gas from the atmosphere. Our planet slowly cooled off from its initial scalding steam bath. After that, the atmosphere would have been made of similar stuff to the air on Mars and Venus: mostly carbon dioxide formed from the reaction of hot rocks with water or oxygen. It's likely to have stayed that way for the next 2 billion years or so, until the proliferation of photosynthetic algae. These precursors of green-leafed plants sucked carbon dioxide out of the sky and in return pumped out vast amounts of oxygen. Our atmosphere is the odd one out principally because the Earth is alive.

Every cloud has a sulphur lining

What would it be like to drift down through the clouds of Venus? From 1972 to 1985, 13 probes – nine of them Soviet – descended through its clouds to steal for us a glimpse of what lies below. Here is what they found:

1972 *Venera 8* (USSR) carried a light-meter to detect how quickly sunlight faded as it descended. From 70 down to about 40 kilometres above the surface, light levels dropped dramatically due to the thick, pervasive layer of cloud. Below that level, despite the fact that there are no clouds at all, you still wouldn't be able to look down and see the surface. The thick atmosphere obscures it, just as the ocean hides the sea floor. But against all expectations, *Venera 8* showed that down on the surface there was still plenty of light. Venus was about as bright as a cloudy winter's day in Moscow.

1975 *Veneras 9* and *10* (USSR) landed and snapped the first black-and-white pictures from the surface. On the way down they detected that the universal thick covering of cloud was, in fact, split into three distinct layers, with clear zones in between. And the clouds were much less dense than those on Earth – not much more than a fine mist.

1978 *Veneras 11* and *12* (USSR) and a cluster of four entry probes from NASA's *Pioneer Venus* mission all arrived in the month of December. The largest American probe had an infrared detector that confirmed the greenhouse effect was entirely responsible for the stifling temperatures. All the probes measured large amounts of sulphur in the cloud layers, confirming the suspicion that much of the clouds on Venus are made of sulphur dioxide – perhaps derived from volcanic eruptions. On Venus it rains sulphuric acid, but that rain boils away long before it ever reaches the ground. *Veneras 11* and *12* also detected a ceaseless stream of lightning bolts.

1982 *Veneras 13* and *14* (USSR) sent back colour pictures from the surface, revealing an orange sky. On Earth the air scatters blue light more than red and yellow, so we see a blue sky. But on Venus the air is so thick that all blue light is scattered away before it reaches the ground, leaving only the reds and yellows.

1985 *Vegas 1* and *2* (USSR) landed and each released a helium-filled balloon to measure wind speeds and weather. At the surface the winds are extremely slow – around 1 or 2 kilometres an hour. Higher from the ground, the balloons were caught in much stronger winds. At one point, *Vega 2* plummeted for several kilometres in a violent downdraft. At much higher altitudes the winds are rushing round at hundreds of kilometres per hour. This 'super-rotation' of the atmosphere above a planet that takes 243 Earth days to spin just once on its axis is still a mystery.

1   *Orange-light zone. Venera 13's view of Venus' volcanic surface.*
2   *Veneras 13 and 14 were some of the largest unmanned probes ever built. The entire lander fitted into the sphere at the top, which is around 2 metres across.*

Getting down

The serene vista of the Earth from space, the curved blurring of sky-blue to space-black, is in reality a formidable barrier to anything trying to enter it. The intense frictional heat generated by slamming into air molecules at several kilometres per second is enough to turn solid rock into ashes, a fact to which meteor trails bear witness. If we were ever to enter the atmosphere of another planet, we would first have to learn how to pass safely through our own.

During the 1950s, American scientists began to experiment with ballistic missile nose-cones. Their first idea was to design a streamlined shape that would, with luck, cut through the atmosphere like a knife. But early studies did not go well: the missiles got far too hot on re-entry. So they opted for a different approach: to hit the air like a hammer rather than trying to slice it apart. The effect of this blunt onslaught was to create a layered flow of air over the attacking face of the missile. This 'laminar flow' was better at keeping the intense heat generated in the air away from the missile's nose, each layer acting as a degree of insulation. This curved disc design went on to be used to great effect in the re-entry vehicles for the first US manned space missions, *Mercury*, *Gemini* and *Apollo*.

The Russians took a slightly different tack for their first probes aimed at the atmosphere of Venus. These robot emissaries were perfect spheres, a shape that protected them with a similar laminar flow to that of the US probes, but gave them less directional control. This didn't matter too much in Venus' thick, soupy air. The first five successful *Venera* probes threw up parachutes to slow themselves down after the heat shield had protected them in the initial burn into the cloud tops. After a redesign from *Venera 9* onwards, the entire globe of the heat shield was discarded on the way down to reveal an hourglass-shaped probe inside. At the top of the hourglass was a flat, wide disc that acted as a brake in the lower atmosphere (which by this point resembled liquid more than air). This slowed the probes enough for them to survive landing using a crumple zone located on the bottom of the lower disc.

1   *Venera 9 is seen here without its encasing heat shield, but the large disc brake is visible at the top.*

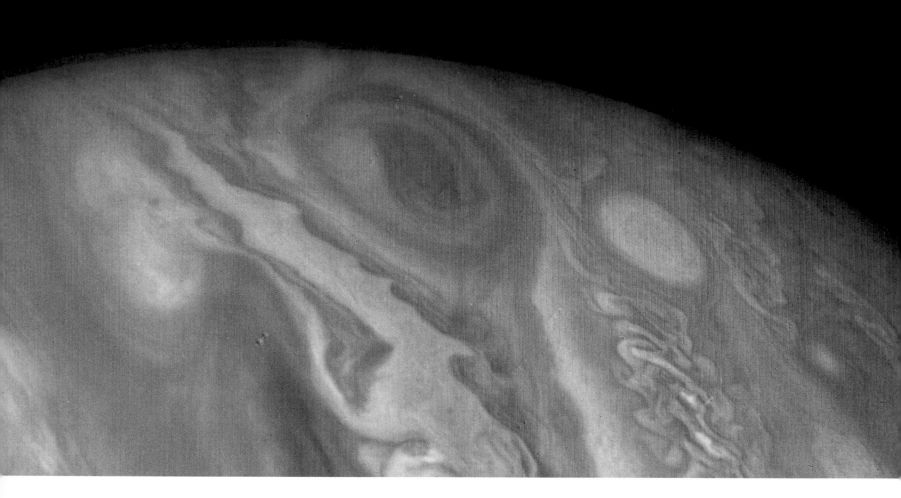

Planet atmosphere

ABOVE *Voyager's view of Jupiter's amazingly complex, turbulent clouds sent mission scientists into rapture. The Great Red Spot was just one of thousands of storms on the planet.*

It's been more than 300 years since the Italian astronomer Gian Domenico Cassini got a clear enough view of Jupiter to see it girdled with those peculiar bands of light and dark, and they have been a source of fascination ever since. As telescopes improved, astronomers would sit captivated for entire nights watching the planet's adjacent belts of russet and cream moving past one another, either with or against the frenetic ten-hourly spin of Jupiter. Also there every night for the past three centuries was the Solar System's most enduring enigma, the Great Red Spot. They would watch it grow and shrink, changing from ruddy brown to bright scarlet. In 1938 three white spots suddenly appeared around the red spot and they've been its faithful companions ever since. But Jupiter is not about constancy; it is about change. There could be no maps of this world. The gas giants are, after all, nothing but atmosphere. Forget Venus. Forget Mars. The place to study extraterrestrial weather was at Jupiter and beyond.

In February 1979, the packed press-room at mission control in JPL was alive with excitement. *Voyager 1* was approaching Jupiter and sending back images of the planet in spectacular detail. The auditorium crackled with spontaneous gasps and ripples of applause. The colours were the first thing to hit them: the swirling streams of ochres, reds, creams and browns. Then there was the red spot, seen in close-up and in glorious colour. It was a massive anticyclone, a storm that could engulf two whole planets the size of Earth. And it was hardly a unique feature; it was just the biggest of 1,000 storms pulsating with lightning as they whipped around this turbulent planet. As more and more detail sank in, the scientists began to feel overwhelmed by the mind-boggling complexity of it all. When the imaging team paraded the pictures of this fantastically complex mélange, its leader, Brad Smith, found himself lost for explanations. His only scientific comment was: 'The existing circulation models have all been shot to hell.'

On Earth the weather is ruled by the Sun. Our bright star heats the equator more than the poles, and that difference in temperature stirs the great mass of the air into motion. Combined with the spin of the Earth, which drags the air round with it, the Sun creates a predictable pattern of weather. The Earth has belts of fairly constant, opposing winds – the trades near the equator and the jet stream at higher latitudes. Where these bands touch, swirling cyclones and anticyclones are stirred up. Seen in that light, Jupiter's weather didn't at first seem that alien. The natural assumption was that the same processes were at work, only the huge mass and rapid spinning of the planet had created more bands of cloud than on Earth.

But ever since 1973, when the infrared detectors aboard the approaching *Pioneer 10* had found that Jupiter generated more heat from within than it received from the Sun, there had been a suspicion that the Jovian weather might instead be a creation of the planet itself, nothing to do with the light that shines upon it. Now *Voyager* added some new evidence: it found that the temperature of the planet was practically the same all over – the poles were barely colder than the equator.

Jupiter's inner glow

In the early 1970s two researchers working in the USA, Frederick Bussé and John Hart, decided to build their own miniature Jupiters to test whether the interior of the planet could create those characteristic bands of cloud. As we saw in Chapter 4, the inside of Jupiter is not solid. It is a ball of liquid hydrogen. Theoretically, a ball of rapidly spinning liquid is supposed to break up into a series of cylinders nested inside one another, each cylinder moving at a slightly different speed from the one next to it. The thought occurred to these scientists that the bands of cloud seen on Jupiter might form at the place where the curve of the planet sliced through the ends of these cylinders, revealing them in cross-section. The belts would then have nothing to do with the Sun, being instead a visible sign of Jupiter's internal structure. Sure enough, their models of the spinning planet, both intricately constructed to simulate the immense pull of Jovian gravity, produced the theoretical cylinders. It was a reality in the lab, but was it so real millions of kilometres from home?

Even after both *Voyagers 1* and *2* had given Jupiter close scrutiny, it was hard to tell which idea was right. Other theoreticians had run computer programs using the Sun as the source of energy and found that they, too, could create weather patterns very similar to Jupiter's. Both camps – internal and external – claimed that the encounters backed their schemes, but neither had conclusive proof.

BELOW *Inside Saturn – an impression of the sight greeting anyone brave enough to sink below the ringed planet's cloud tops.*

While the arguments raged, *Voyager 2* sailed on through the Solar System regardless. Saturn's bands were subtle, but essentially a muted replica of what had been seen at Jupiter. Again, both sides argued their points with equal vigour. Then, at Uranus, came the tie-breaker.

Uranus, as we know, is tilted on its side. In 1986 its south pole was pointed almost directly at the Sun and the north pole had been in darkness for 20 years. If the solar heating camp were right, the weather on Uranus should be nothing like what had been seen on Jupiter and Saturn: in fact, the winds should have been streaming away from the south to the north. At first, no one could tell what was happening. The pale aquamarine orb initially showed no features to *Voyager*'s camera. But then, with the image enhanced to bring out the most contrast, and with the help of the infrared readings, an answer materialized. The bands on Uranus were there, just as they were on Jupiter and Saturn. In fact, the equator was marginally the hottest place on the planet. Bussé and Hart felt completely vindicated. These planets didn't need the Sun – they made their own weather.

As if to ram the point home, Neptune, where the heat from the Sun is only one-thousandth of that we receive on Earth, has some of the most exciting weather in the Solar System. *Voyager 2* saw wispy white clouds appear and disappear within minutes. Even the Great Dark Spot that stunned the *Voyager* scientists in 1989 had disappeared by 1996 when the Hubble Space Telescope turned to the most distant giant planet. And Neptune's changeable clouds confirmed a trend that had first been noticed back in 1980 at Saturn: the speeds at which the bands move around the giants increase the further you travel from the Sun. How could that be unless the planets themselves were driving the winds?

But that was not the end of it. Jupiter is the closest to the Sun and receives the most heat from it. Traditionalists still argued that the Sun could be driving the weather on Jupiter. To know for sure scientists would need to delve deeper into this gassy world, to get down into the atmosphere and see what things were like away from the glare of daylight.

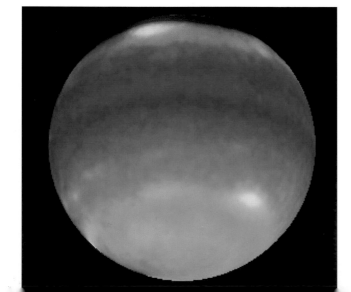

Wild worlds
of weather

If you're planning a trip around the Solar System, make sure you go prepared for some weird weather. First stop is Mars, where the onset of winter brings a chilling sight to planet's south pole. As the sky overhead cools to a state of unpleasant frigidity, a fine snow of carbon dioxide begins to fall. Moving on to sweltering Venus, chances are you'll catch a very different type of snow – made of pure metal. Down in Venusian valleys the temperature is hot enough to evaporate some metals, and having spotted bright caps on top of some high mountains in images from the Magellan probe, a few scientists think these slightly cooler regions might be covered in a metallic snowcaps. On Saturn's misty moon Titan, there's another sight to perk up the weary traveller: raindrops of liquid methane. Floating slowly down in one-sixth the gravity of that on Earth, they should have time to grow into droplets up to a centimetre across – twice the size of Earthly rain.

All these sights pale, however, next to the grandeur of the swirling atmospheres of the gas giants – foremost is Jupiter's Great Red Spot. This blotch on the giant planet has been observed through telescopes for 300 years, and could well have been around for much longer. How can a storm, something that lasts a few days at most on the Earth, keep going for so long?

The *Voyager* probes had the chance to study Jupiter's weather close-up during two 30-hour 'cloud watches' in their 1979 flybys. The red spot and several other large storms carried on as usual, but the cameras saw thousands of small cloud swirls being created and destroyed. The probes discovered an important difference between small storms and their larger cousins. The small storms form when they are sheared off the edge of one cloud band by an adjacent band travelling the other way. Within one or two days, however, the opposing forces of the two bands pull them apart. But the large storms seem to survive by rolling with the currents – they allow the cloud bands either side of them to spin them around.

How this really works is a mystery, although theories abound. The red spot and the other long-lived storms appear to 'eat up' small storms, and some scientists think this is what keeps them spinning at the top of the atmospheric found chain. Others think they are kept wound up by some supply of energy from below, deep down inside Jupiter. Or it could just be that on a planet the size of Jupiter, where everything is writ large, it simply takes more than 300 years for a storm to blow itself out.

1 The storm that goes on and on: Jupiter's Great Red Spot.
2 The south pole of Mars is covered with frozen carbon dioxide.

Rendezvous with a gas giant

The idea of sending a probe into Jupiter was not a new one. Back in the early days at NASA Ames, many of the original mission concepts for the *Voyagers* involved plans to send craft that would drop probes into each of the gas giants. Veteran engineer Alvin Seiff recalls that, at the time, he pointed out that Jupiter would be the hardest planet to enter safely. This is because when a spacecraft reaches a planet it is accelerated by that world's gravity. On Earth any arriving probe will reach at least 11 kilometres per second when it hits the atmosphere. On Jupiter the minimum arrival speed is 60 kilometres per second – 210,000 kilometres per hour!

In the late 1970s Seiff, along with a team of engineers, got the go-ahead to design the ultimate probe. He had now designed nose-cones for spacecraft that had penetrated the atmospheres of Mars and Venus, but the Jupiter entry probe would be the toughest test of human engineering. First there was the entry itself, and then there were the mounting pressures of the planet to contend with as the probe fell gently inwards. At the centre of Jupiter the pressure would be millions of atmospheres. The probe had no chance of making it to the heart of the planet, but it had to make it far enough down into the atmosphere to get away from the heat of the Sun and solve the mystery of the winds. The team opted for a vented design; Jovian air would be able to flow freely in and out of the probe to eliminate the problems of pressure, but that meant the rising temperatures would be its demise instead.

Seiff would never have believed that he'd have to wait so long to see if the design would pay off. The *Galileo* mission, which would carry the probe on the back of a Jupiter-orbiting spacecraft, was plagued with setbacks. NASA policy changes, launch delays and, most of all, the crash of the space shuttle *Challenger* in 1986 conspired to make the mission arrive at Jupiter ten years behind schedule. Nevertheless, in July 1995, the *Galileo* orbiter began to feel the draw of the most massive planet in the Solar System.

That was the cue for it to release a barrel-sized probe and send it drifting headlong towards the swirling psychedelic mass of cloud below. As the *Galileo* probe became the first man-made object to touch the atmosphere of a gas giant, it also became the fastest thing ever built. At the moment of entry it experienced a scorching braking force 228 times as strong as Earth's gravity. For 30 seconds the tiny probe weighed as much as a jumbo jet and its heat shield soared to nearly three times the surface temperature of the Sun. But despite being in uncharted territory and a billion kilometres from home, *Galileo* was not alone. Back on Earth hundreds of scientists were tracking its every move. Through telescopes and satellite dishes they watched and listened, hoping and praying it would survive. Soon *Galileo* had slowed to 100 kilometres per hour and 60 minutes of the most spectacular descent imaginable lay ahead: a leisurely fall through the upper atmosphere of the most complex and fearsome of all the skyscapes in the Solar System.

PAGE 185 *The windy planet: Voyager 2's close-up of wispy clouds streaming along in Neptune's 1,500-kilometre-per-hour winds.*

A tantalizing glimpse

The data sent back in that hour was to crack open the door to a new world. Apart from the winds, the *Galileo* probe returned long-awaited news of the chemical make-up of Jupiter, the temperatures and pressures inside the gas giant, and the nature of the clouds in that strange sky. At the time of writing, scientists are still wringing the last drops of information from the probe's short stream of data.

The descent was a victory march for Bussé and Hart. As the probe fell, it watched the wind speeds rise, from 400 kilometres per hour at the cloud tops to over 600 at 40 kilometres in, and then remaining steady all the way to 130 kilometres down, when the little probe succumbed to the 150-degree Centigrade heat. Only if the weather on Jupiter were driven from within would the wind speeds increase as the Sun's intensity faded behind the clouds.

And what of those clouds? Scientists expected Jupiter to boast three separate layers of clouds: ammonia the highest, next ammonium hydrosulphide, then water. *Galileo*'s kamikaze route should have taken it through all three, but a special device designed to detect particles suspended in the air — a nephelometer — found only one wispy region. It was most likely a puff of ammonium hydrosulphide, but no water or ammonia clouds materialized. It turned out that the probe had headed straight for an unusually dry spot on Jupiter, where hot air from deep in the planet rushed up to the surface. The mission scientists soon started referring to it as the 'Sahara desert of Jupiter', and they spotted by telescope several other small areas like it around the planet. It was a piece of incredible bad luck. What the vast majority of Jupiter's atmosphere is like below the cloud tops remains a mystery.

One more place to look

It will be a long time before we have the chance to see inside a gas giant again. But there's another atmosphere on the horizon. Circling Saturn is the veiled satellite Titan. It is the only moon in the Solar System to have a substantial atmosphere, and with unbroken cloud cover reminiscent of Venus, it is arguably the most intriguing world in the Solar System. A place that by rights shouldn't have any air at all boasts a covering of gas thicker than the Earth's. It is the only atmosphere in the Solar System, aside from the Earth's, to be made predominantly of nitrogen. The ubiquitous reddish-brown haze seems to be made of organic chemicals similar to those found in car exhaust. What could it be like below the freezing, dimly lit clouds of this hidden moon? On 27 November 2004 the world will find out when the European Space Agency's *Huygens* probe dives in. And this time there will be an innovation. All the way down on its two and a half hour journey to the surface, a camera will snap pictures of the alien scene around it. For the first time we will see what it is like to fall through the sky of another world.

OPPOSITE *The* Galileo *probe makes its way out of the cargo bay of space shuttle* Atlantis *on 18 October 1989. The probe that would pierce Jupiter's clouds six years later is sandwiched between the main* Galileo *orbiter and the white booster rocket.*

ABOVE *An impression of the cloud tops of Saturn's moon Titan. What will the* Huygens *probe find beneath this chilly blanket when it reaches Titan in 2004?*

life

7

THE PROBE SITS ON THE RED soil, watching, waiting. For eight days now, *Viking 1* has been bringing the world the first breathtaking vistas of Mars, but its real mission is only just beginning. Now a long arm telescopes its way down to the alien soil, a small scoop digs at a bare patch next to a rock recently named Shadow. Scientists on Earth wait blindly, hoping that the rest of the robot's contortions will choreograph as they have so painstakingly planned: the 18-minute time difference makes real-time communication with the craft out of the question. The scoop now retracts and swings over a small hopper sitting on top of a metal cube measuring about 30 centimetres on each side. The soil trickles down into the hopper, which then starts a journey around a small circular track, delivering like manna small doses of Martian soil to each of the three experiments crammed into the box. The soil is what the scientists have been waiting for all those long months of flight, and the little probe immediately gets to work on it. Back in mission control somebody screams, 'The PR light is on!' The experiment to test for photosynthesis has begun. This is the moment of reckoning. Science is finally asking the question that has haunted us for centuries: is there life beyond the Earth?

three days later, on 31 July 1976, Chuck Klein, the *Viking* project's lead biologist, was standing in front of a throng of journalists and TV cameras. He said he had 'important, unique and exciting things' to tell. When they heard what he had to say, the reporters went into a frenzy. News bulletins that night declared that there really could be life on Mars. NASA hadn't had so much attention since the *Apollo 11* manned Moon landing seven years earlier. Only two of the three different tests for life had reported back, but their results were so emphatic that Klein opted to go public before the remaining one came in.

PAGES 190–1 *The Sun glinting off the only world known to possess a life-incubating layer of liquid water: the Earth.*

Klein was not announcing anything quite as dramatic as the discovery of little green men. Although mission scientist Carl Sagan had given himself the task of scrutinizing *Viking 1*'s photographs for signs of movement on the Martian surface, by the time of Klein's press conference he had already given up, quipping, 'So far, no rock has obviously got up and moved away.' Klein's goal was less easy to track down. He and his team of biologists were interested in the possibility that microscopic life might exist on this cold and barren planet, and they had spent years trying to design a foolproof set of experiments that could root it out.

It was still 28 July when the first stunning results came back. Two and a half hours after the soil had dropped into his soil sampler, Vance Oyama got readings he couldn't believe. The biochemist from NASA's Ames Laboratory in California had built an experiment to squirt a

ABOVE LEFT Viking 1 *took this photo soon after it began to test for life in Martian soil.*

ABOVE RIGHT *The* Viking 2 *probe pushes a rock aside to get to the precious soil below. 'Would it contain Martian microbes?' was the question in scientists' minds.*

ABOVE *Hold still: the camera on the flight spare of the* Viking *lander practises on project members at NASA's Jet Propulsion Laboratory in Pasadena, California. The* Viking *cameras recorded images very slowly, one vertical scan at a time – moving from left to right, it took so long for the camera to take this picture that scientists on the left were long gone while those on the right were still posing.*

chemical 'chicken soup' on to the Martian soil. If any bacteria were alive in there and they liked Earth food, Oyama hoped they would give off some gas, and that was what his gadget was looking for. Martian bugs weren't expected to be hyperactive, eking out a living on such a cold planet, and Oyama had been prepared to wait weeks, even months, for anything to happen. In the event, he hardly had to wait at all. Those first readings showed a huge amount of oxygen had risen out of the soil after just a teaspoon of his alien broth. The soil appeared to be teeming with life.

Then, at 7.30 p.m. on 30 July, *Viking* started sending back the results of the second experiment. They came back to Earth in the form of a long computer printout which appeared so shattering that Gil Levin, the experiment supervisor, had his team solemnly sit down and sign the first page. The 'labelled-release experiment', as it was called, was similar to Oyama's test, but Levin's 'soup' included a radioactive form of carbon. If his broth were digested by Martian microbes, their satisfied belch of radioactive gas would reveal itself to the on-board Geiger counter. The printout showed clearly what had happened. 'The odds were overwhelming that nothing would happen at all,' said Levin, 'and when we saw that curve go up, we flipped.' Almost immediately after the nutrients were added, the Geiger counter had gone right up the scale.

Second thoughts

Although Levin and Oyama were faced with results that before launch they would have wasted no time in calling positive, it wasn't long before many on the biology team were having second thoughts. Martian soil seemed more active than even the most fertile soil on Earth. How could the reaction be so fast? Could it be that the reactions weren't biological at all; might there instead be something else in the soil that was causing a release of gas? Three days after Chuck Klein's press conference and nearly a week after the third experiment had started, *Viking* reported back: it, too, apparently proved positive. Biologist Norm Horowitz had designed his pyrolytic release (PR) test to look for photosynthesis, a sign of primitive Martian plants. Horowitz also used radioactive carbon, but this time in the form of a gas, carbon dioxide. After the soil had been exposed to gas and light for a few days, the soil, or something in it, appeared to have 'fixed' some of the radioactive carbon, just as plants would have done back on Earth.

By now the biology team was at fever pitch. Levin was clinging fast to his positive result as proof of life. But another possibility was weighing more and more heavily on the rest of the team. The presence of a reactive, corrosive chemical in the soil might account for all three positives. Something like hydrogen peroxide, a form of bleach, would mimic all the reactions recorded thus far, and it would act quickly, too. The three tests couldn't resolve the issue by themselves. Klein looked to a fourth *Viking* experiment, one from outside the biology team, to settle the issue.

Back on 28 July some Martian soil should also have been delivered to a device that would look for organic chemicals. Organics are molecules made from carbon and hydrogen – sometimes extremely large and complex – and they are the fundamental building blocks of all life that we know. If there were life on Mars, it would surely leave organic traces of itself in the soil. But something went wrong as the scoop went back for its second dig at the red earth. The light back at the Jet Propulsion Laboratory (JPL) that was supposed to come on when the test chamber felt the weight of the soil tumbling in didn't illuminate. Because of the flurry of commands being sent up to *Viking 1*, the final organic chemistry experiment had to wait another six days before it could be retried. But the test appeared to be jinxed and the scoop got jammed on the way to making its delivery. By this time, the organic chemists were going crazy, and even though the weight gauge was still reading zero, they had seen a small pile of soil right next to their hopper on one of the *Viking* photos. They gambled that some soil – not enough to register on the gauge but enough to run the test – had got in and they decided to go ahead. It was a big risk: the equipment had cost $55 million (equivalent to $165 million today) and if there were no soil in the container, they would waste the test. But the gamble paid off…to devastating effect. The results said without a doubt that there were no organic chemicals in Martian soil – it was absolutely lifeless.

The *Viking* biologists had nothing left to argue about: the positive results must have been caused by a reactive chemical in the soil. Somehow, through aeons of baking in the intense ultraviolet light from the Sun, which passes easily through the thin Martian atmosphere, the soil had become viciously corrosive, full of chemicals that would burn your flesh if you dared to step in it. The world awoke abruptly from its long dream of life on Mars.

ABOVE *Engineers work to miniaturize all three complicated sets of experiments to test for life on Mars into a box measuring a mere 30 centimetres wide. The kit was only just completed in time for the* Viking *launches.*

Norm Horowitz, who had been so confident before the landing that there would be no life in Martian soil that he'd boasted he would gladly eat a spoonful of it on a salad, withdrew his offer, but only on the grounds of the corrosive peroxide. Gerald Soffen, the *Viking* project scientist who had put 15 years of work into the search for alien life, left JPL a disenchanted man after the landings. He went on a teaching sabbatical at Harvard, and on day one of his classes he wrote on the blackboard, 'GOODBYE MARS, HELLO EARTH'. Mars, he said, was a planet for geologists, not biologists.

The age of dreams

How different our ideas had been before the space age. At the end of the 19th century life on Mars had been respectable talk in the gentlemanly society of science. When the Guzman prize was announced in Paris – a staggering 100,000 francs for the first person to make contact with an extraterrestrial species – contact with Mars had been specifically excluded from the competition. Talking to Martians would be far too easy, it was thought, since their existence was all but a certainty.

The Victorian popularity of Mars owed a lot to the craze for Martian canals inadvertently stirred up by Italian astronomer Giovanni Schiaparelli (see Chapter 3). But the idea that there might be other forms of life in the Solar System dates back at least two centuries earlier, to Dutch astronomer Christiaan Huygens. If Galileo and Copernicus realized that there was nothing special about the Earth's position in the Solar System, Huygens took the next logical step:

ABOVE *Dutch astronomer Christiaan Huygens (1629–95) was the first scientist to declare his belief in the existence of aliens.*

A man that is of Copernicus' opinion, that this Earth of ours is a planet carried round and enlightened by the Sun like the rest of them, cannot but sometimes have the fancy…that the rest of the planets have their dress and furniture, nay and their inhabitants, too, as well as this Earth of ours.

Christiaan Huygens, *New Conjectures Concerning the Planetary Worlds, Their Inhabitants and Productions*, c.1690

Huygens asked the question that we still ask today: what is so special about the Earth? If ours was just one of many planets going round the Sun, could there not be other places in the Solar System where life had taken hold? It was a dream that fired up the pioneers of the space age. Most of the scientists who designed and built the first interplanetary spacecraft had grown up reading the science fiction of H.G. Wells and Edgar Rice Burroughs – stories of princesses on Venus, or of the travels of Virginia gentleman John Carter to Mars. It didn't take long to kill those dreams.

The myth of Venus was the first to go. As we saw in Chapter 6, radio signals from the planet hinted that its temperature might be hundreds of degrees Centigrade, and the early

robotic missions confirmed this was a boiling, corrosive cauldron of a world. It is odd now to think that the first Venus probe, *Mariner 2*, was hastily sterilized before launch just in case it crashed into the planet and contaminated it with earthly microbes. Many sterilization processes on Earth would fail to eradicate life as efficiently as the pressure-cooker atmosphere on Venus.

But NASA was still prepared for life in the Solar System in the late 1960s. Astronauts and rocks returning from the Moon were placed in quarantine, in case they contaminated the Earth with lunar microbes. In the end, the fears were unnecessary. Tests showed that not only was there no life on the Moon, but that virtually all its soil was utterly dry and had been completely sterilized by the Sun's ultraviolet radiation. (NASA's *Lunar Prospector*, launched in 1998, has recently found evidence for tiny amounts of water ice hidden in permanently shaded craters at the Moon's poles, but the vast majority of lunar dust is drier than any desert on Earth.)

Jupiter was the next to fall. Arthur C. Clarke, science fiction writer and inventor of the idea of communications satellites, had popularized the notion of life forms floating around in the cloud tops of Jupiter like hot-air balloons. As late as 1969, after he had learnt that Venus was unbearably hot, he declared that Jupiter must be considered the best bet for life in our Solar System. But the pair of *Voyager* craft and, more recently, the *Galileo* orbiter found the upper atmosphere of Jupiter to be incredibly turbulent. It is highly unlikely that any forms of life could survive the extreme changes in temperature and pressure, as they would be one moment deep in the crushing, scalding depths of the planet's atmosphere and the next soaring high into its rarefied and freezing cloud tops.

And then *Viking* killed Mars. Beyond the confines of the Earth, life, it appeared, would be fried, irradiated, desiccated or corroded. Huygens' dream was dead. Space was beginning to look like a desolate place.

Another Goldilocks?

BELOW *The closest view we have ever had of Jupiter's moon Europa. The* Galileo *orbiter swooped just 560 kilometres above its frigid, cracked surface. The terrain looks strikingly like the Antarctic on our own planet.*

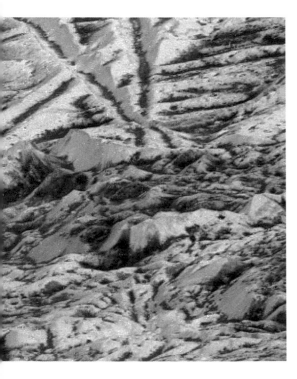

BELOW *The closest view we have ever had of Jupiter's moon Europa. The* Galileo *orbiter swooped just 560 kilometres above its frigid, cracked surface. The terrain looks strikingly like the Antarctic on our own planet.*

When *Voyager 1* approached Jupiter in 1979, no one on the team gave any thought to life. There was not a single biologist involved in the mission. Instead, the scientists prepared to meet four giant gassy worlds and to study the hordes of satellites that whirled around them. Nobody expected the moons to grab any headlines. Brad Smith, *Voyager*'s imaging team leader, had expected to find nothing more than inert lumps of icy rock bearing the scars of billions of years of pelting by meteorites. In the event, only Callisto met those meagre expectations. Jupiter's three other large moons ended up stealing the show. Io was top of the bill, turning out to be a world chock-full of active volcanoes. Ganymede, the Solar System's biggest moon, boasted a bizarre grooved surface that looked like it had been raked over while the ice was still freezing. But Europa was the biggest mystery. Its icy face, almost completely smooth, must have been recently melted. But how?

An unexpected force is at work inside these moons: tidal heating. As we saw in Chapter 3, the colossal gravity of Jupiter is wrenching at their cores as they dance around the giant, creating enough heat to melt their insides. On Io, closest to Jupiter, the heat is strong enough to melt rock and create a volcanic hell. Ganymede, the furthest out of the three, had received only the gentlest gravitational warming at some time in the past, leaving those strange markings on its now frozen body. Europa is slap bang in the middle – not too hot, not too cold. Although no spacecraft has yet been able to prove it, everything points towards conditions being just right for the key ingredient to life: liquid water.

The chance that there might be an ocean of liquid water lurking below Europa's icy crust filled biologists with new hope. They had come to view Earth as the 'Goldilocks Planet'. Venus was too close to the Sun and too hot for water; Mars, further away, was too cold. But Earth was just right. Liquid water seemed to be the one thing that set us apart from the rest of the planets, and as long as the Earth was the only water world, perhaps it would prove to be alone in harbouring life. Suddenly, though, it seemed there might be another ocean in the Solar System, a sea that no explorer has yet navigated.

The new breed

If the hunt for life got a boost at Jupiter, Saturn wasn't about to be outdone. Its moons are a clutch of small icy worlds for the most part, but one always stood out. Titan was the second biggest moon in the Solar System and the only one to have a substantial atmosphere. Looking from the Earth, astronomers knew that its air contained methane, or natural gas. On Earth natural gas comes from living creatures, either when they are alive or after they die. No one thought this was the case on Titan – at 200 degrees below zero, it's far too cold for life. But could Titan's methane be a sign of more complex organic chemicals; in other words, might this world hold some of the basic building blocks of life in a cosmic deep freeze?

In November 1980 *Voyager 1* reached the domain of Saturn, and one of the key mission objectives was to take a closer look at Titan. It skimmed by just 4,000 kilometres above the clouds, glimpsing nothing of the surface beneath the thick atmosphere. But its other instruments found plenty for the chemists to mull over. The ball of gas around this moon was mostly nitrogen, just like the one around the Earth today. But there was also a dash of methane, ethane, hydrogen cyanide and other chemicals which together looked disarmingly like the primordial soup which, it is thought, fostered life on Earth billions of years ago.

Titan and Europa were two bright beacons of hope. Titan proved that the stuff of life was spread across the Solar System. Europa showed us we'd been too blinkered – there might be other places around our star where life could get a foothold. Suddenly, a new type of scientist emerged – the exobiologist. How could life survive in the extremes of outer space? Can we find examples of life on Earth that are tough enough make it on Europa, or perhaps on a corner of Mars not touched by *Viking*? That's what exobiologists were about – the biology of the beyond.

ABOVE LEFT *Frigid air: beneath its thick cloud, Saturn's moon Titan is believed to contain the frozen organic building blocks of life.*

ABOVE RIGHT *This area of Europa's icy surface looks like it may have been melted and flooded by an ocean below it, before freezing over again.*

To the ends of the Earth

One night in 1978 Imre and Roseli Ocampo Friedmann were sitting watching television at home in Tallahassee, Florida. Just in from a day's work at the university, they tuned in to the evening news and were stunned to see that they'd hit the headlines. Two years after *Viking* had seemed to close the book on Martian life, eminent TV anchorman Walter Cronkite was telling the nation that the discoveries of this pair of microbiologists were giving the possibility of life on the red planet a second chance. The Friedmanns were shocked to see themselves on the news, but in a sense it was high time. All along, Imre Friedmann had said that *Viking* was looking for life in the wrong places. Imre knew that the place to look for life was not in the soil, but hidden inside rocks.

ABOVE LEFT *A slice of life: a cut through this sandstone rock from Antarctica shows a thin green layer just below the surface, which is teeming with microscopic life.*

ABOVE RIGHT *Cryptoendoliths: a close-up of these hardy rock-dwellers surviving where the temperature barely ever rises above freezing.*

The story had begun back in 1961 when, after spending ten fruitless years searching for algae in Israel's scorching Negev Desert, the Hungarian-born scientist had a rock pressed into his hand. His university colleague, a geologist, had been searching for oil when he broke open a piece of limestone and found inside a curious green layer. 'He told me that he'd thought at first it was copper, but it wasn't. So he gave it to me because it might be biological,' recalls Imre.

When he tested the green layer, it really did turn out to be alive: it was a type of blue-green algae living off small pockets of water trapped in the rock. It survived in seemingly impossible conditions because the rock could hold water for months, even when the weather outside was dry as a bone. After that, Imre found algae hiding in rocks all over the desert. He began searching for porous rocks – sandstones and limestones – looking for telltale mottled markings on their surfaces. Chipping away a few millimetres of the rock, he would often find a

LEFT *Mars on Earth: the bitter cold and lack of rain or snow makes the Dry Valleys of Antarctica a veritable test of the tenacity of life.*

BOTTOM LEFT *Some scientists think life could exist around underwater thermal vents on Jupiter's icy moon Europa, just as it does around 'black smokers' under the Pacific.*

BELOW *The water spurting from this geyser in New Zealand is nearly boiling, but microbes proliferate in the scalding ponds nearby.*

What is life?

Life is something that we all instinctively think we would recognize here on Earth, but how easy would it be to recognize alien life forms? All life we know of is built from carbon-based chemical compounds. Carbon atoms are so good at joining together into long chains and pulling in other elements that an entire branch of chemistry has been named after them. Organic chemicals are complex groups of atoms that we call by simple names such as protein, sugar or acid. There are millions of possible organic chemicals, but life on Earth uses only a tiny handful of the full repertoire. Is it necessary for life elsewhere to use the same small selection, or could alien organisms be less choosy? There is even the chance that living entities could be based around the chemistry of an element other than carbon. Silicon has often been suggested as a good runner-up to carbon. Life in silicon-based clays or rocks is entirely feasible, as are other possibilities such as life in freezing liquids or based on the chemical ammonia. Clearly, expecting to find alien life in the form of cells with genes made from DNA and filled with familiar proteins is too narrow-minded.

Perhaps it is better to try to define life not by what it is made of but by what it does. Charles Darwin would have said that life is something that mutates and evolves from generation to generation. But is it necessary for life to reproduce? There seems to be only one definition that life can't duck out of – the need for energy. All life must use energy to keep going, and in this universe full of trillions of nuclear power stations called stars there is no shortage of that.

1   Deoxyribonucleic acid, or DNA, is the foundation of life on Earth. Its double helix contains all the information needed to make a bacterium, a mouse or a human. But life elsewhere may not be based on DNA at all.

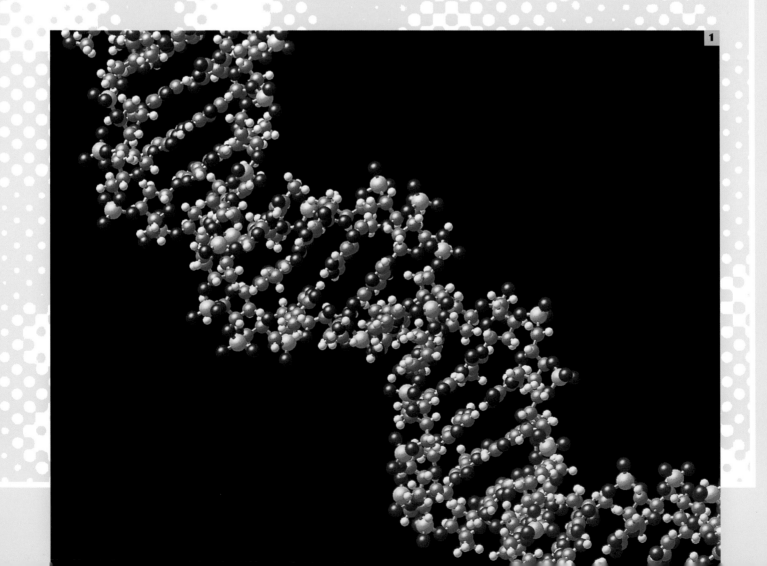

rich green layer lurking below. Imre named the new organisms 'cryptoendoliths', meaning 'things that hide inside rock'.

Imre's thoughts soon turned to finding life in even more inhospitable environments – places that were not only dry, but freezing cold too. However, his requests to travel to Antarctica in search of life were met with ridicule. His work in the Negev had received little publicity. In any case, NASA had spent much of the 1960s searching for life in the Antarctic Dry Valleys, testing the prototypes of the experiments that would one day fly on *Viking*. If their state-of-the-art experiments had drawn a blank, they weren't about to listen to some newcomer with a crazy idea about life in rocks.

In the summer of 1973 Imre's luck seemed to change. He met Wolf Vishniac, a biologist working on the *Viking* mission, whose own life-seeking experiments had just been dropped to cut down on spacecraft weight. Imre's rejection by NASA must have struck a chord with Vishniac, who, as fate would have it, was about to leave on a field trip to Antarctica. Vishniac promised to bring him back a rock sample from the South Pole.

Imre remembers hearing the news of the tragic accident only too well. A few months into his field trip, one night just before Christmas, Vishniac had gone out to check his equipment on the slopes of Mount Baldr. His body was discovered the next day: he must have slipped and fallen. For Imre it was a double blow. He had lost a new friend and a rare ally; and his one chance to search for life inside rocks in Antarctica had gone with him.

In March 1974 Imre received a letter from Helen Vishniac. She had just had her late husband's belongings shipped back to her, and among the cases she had found some rocks that had Imre's name on them. Imre and Roseli couldn't wait to get them back to the lab. In a bittersweet moment, they discovered a perfect specimen of blue-green algae inside one of the rocks. Here was an organism making its home quite happily in a place where the temperature rises above freezing no more than one or two days each year. They had proved that life on Earth was much more resilient than anyone had ever believed.

Their discovery was momentous, but the timing was terrible. The Friedmanns' discovery of life in Antarctica hit the scientific news-stands just three weeks after *Viking 2*'s metal footpad touched the red dust of Mars. The papers were still full of the possibility of life on Mars – life on Earth, no matter how unlikely, wasn't considered newsworthy. It was over a year later, when the space programme was suffering from post-*Viking* blues, that the Friedmanns heard that NASA wanted to do a press release on their work.

In several ways, 1978 marked a watershed in ideas about life on Earth and the possibility of life outside it. The discovery of life in the Dry Valleys of Antarctica, a place not so different from Mars, showed just how tenacious life could be. And that same year, 2,400 metres down on the ocean bed, two men in a deep-sea submarine were stunned to find life flourishing around a smoking volcanic vent near the Galapagos Islands. Until that moment, it had been a tenet of biology that life on Earth was fed by sunlight. The microbes living there needed no sunlight – they had never seen it. Instead, they used the hot gases of the Earth to survive.

BELOW *Wolf Vishniac stumbled across the first examples of life in the Dry Valleys, but the Antarctic claimed his life before he could bring back the samples.*

All you need is water

BELOW *Waterworld: Viking orbital photos show that Mars of old was a world full of rivers and valleys. Mangalla Valles, on the right, lies next to a flat plain that might once have been an ocean bed.*

The rapidly expanding science of exobiology has already consigned *Viking's* lifesearch to ancient history. In the past quarter century we have learnt that life on Earth can exist in all manner of extreme environments – without sunlight, where it's freezing cold, where it's boiling hot – as long as there is water. None of these life forms is very complicated, none of them is bigger than a single cell, but it's life all the same. For that reason, the search for life in the Solar System has become the search for water. Find water, then check for life. Earth is the only world

that we know for certain to have an abundance of liquid water. Today, Europa alone might join our ranks. But what about in the past? What was going on billions of years ago, when the Solar System was still young? Did the Earth have the monopoly on water back then?

As we saw in Chapter 6, the Earth and Venus shared very similar origins. Both planets formed as balls of molten rock surrounded by a thick veil of steam. On the Earth, the planet slowly cooled and the steam rained down to form the oceans. Venus might also have had a brief period with an ocean, but being closer to the Sun, it was always much hotter and any water would probably have been close to boiling away, which it eventually did as its greenhouse effect ran out of control. Could there have been a brief honeymoon for life on Venus, clinging fleetingly at the shores of a warm sea? We will never know because the searing temperatures under Venus' veil have long since wiped clean any trace it might have left.

Mars has a different history, and it's written all over its face. In 1971 *Mariner 9* went into orbit around Mars and discovered that this was a world with a fascinating past. It saw enormous volcanoes and giant chasms for geologists to drool over, but it also found something for the biologists. It was clear that there had been vast amounts of water here. *Mariner* glimpsed huge outwash channels carved by flash floods that must once have discharged half a billion tonnes of water each second — 1,000 times the flow disgorged by the mouth of the Mississippi.

It is ironic that just as the *Viking* landers were writing off the idea of life on Mars, the pair of orbiters that had released them were continuing the photographic reconnaissance of *Mariner 9* and amassing evidence of a planet that might once have been a perfect haven for life. The southern hemisphere of Mars abounds with vast networks of valleys and intricate webs of channels and tributaries that could only have been carved over the aeons by running water. Judging by the number of craters on them, the valley systems had to be nearly 4 billion years old. In 1976 it was widely thought that this meant water hadn't lasted long enough on Mars for life to take hold. Today, we're not so quick to write off that possibility.

Mission to planet Earth

At the time of writing, mankind has failed to find any life beyond the Earth. But what are our chances of doing so in the future? We know that there are places where life could possibly exist and places where life might have been in the past, but how likely is it that life actually got started anywhere? To answer those questions we have to delve into the mystery of how life began on Earth. Was it easy — 'just add water and wait' — or did it require a series of incredible coincidences on a planet where all the conditions were perfect? Perhaps if we could find out how soon life took off on Earth, we might know what the odds were of it taking hold on Mars, Europa and elsewhere.

Humans, fish, birds, insects, dinosaurs: all the life you can imagine represents only the blinking of an eye on a geological time-scale. For most of the history of this planet, nearly the first 3 to 4 billion years, there existed almost nothing more complicated than single-celled protozoa. So the hunt for the first traces of life is unimaginably harder than digging up the

bones of an ancient ape or a *Tyrannosaurus rex*. But, remarkably, some of these ancient microscopic fossils have actually been found.

Working in 1991 in western Australia, American palaeontologist Bill Schopf unearthed a biological treasure hidden in ancient sedimentary rocks. In among the dried-out remains of a 3.5 billion-year-old pond, he stumbled across tiny, fractured worm-like forms, the remains of primitive microscopic algae. These are the oldest fossils ever found and they bear more than a passing resemblance to similar organisms that are still alive today. Clearly, life had advanced quite a bit when the Earth was only a billion years old. It seemed likely that there would be older fossils lurking somewhere. But where to look? We live on a planet that is forever reshaping its continents and oceans, where volcanoes have smothered ancient fossils with lava, and tectonic plates have dragged mummified microbes underneath its crust for billions of years. This relentless geological activity makes it unlikely that any hunk of rock would survive unscathed from the time it was first formed. Could it be possible to find traces of fossils in older rocks damaged by the Earth's geology?

Steven Mojzsis, a young researcher working at the Scripps Oceanography Institute in San Diego, California, set out in 1995 to look for the microscopic remains of life in rocks more than 3.5 billion years old. There aren't many places on Earth where you are likely to find rocks that ancient. Australia is one place, and the icy wastes of Greenland are another. Mojzsis spent one Arctic summer combing the shores of Greenland, chipping out and collecting the rocks he thought looked the oldest. He chose sedimentary rocks, seams that formed from material sinking down through pools of water, since he imagined that life would be most likely where there had been water. Several months later, the rocks finally arrived back in San Diego in huge packing crates. He hastily prised them open with a crowbar and began chipping away at the samples. In them he discovered something that has changed the way we think about the origin of life in the Solar System.

ABOVE *Everyone's great, great grandmother: these microscopic fossils discovered by Bill Schopf in Western Australia are similar to modern-day pond scum and are 3.5 billion years old.*

RIGHT *Rock graveyard: some of the oldest rocks are found in the icy wastes of Greenland, including some sedimentary rocks that formed at the bottom of the Earth's earliest ocean.*

Mars of old

In 1894 a wealthy Bostonian journeyed west to the high desert of Arizona to build an observatory. Percival Lowell's imagination had been fired by Italian astronomer Giovanni Schiaparelli's observations of '*canali*' on Mars. The Italian had meant simply to describe his vision of thin channels, but this was mistranslated as 'canals'. Lowell took that at face value and sat for years at his 24-inch telescope mapping the intricate web of artificial waterways that appeared before his eyes. So keen was he to keep up the vigil that sometimes he would have to dangle from a ladder halfway up the dome to catch a glimpse of Mars when it was low in the sky and the telescope's eyepiece was correspondingly many feet off the ground. He made countless sketches of his beloved canals.

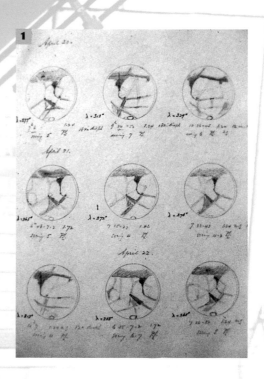

Lowell had a passion for Mars, and a literary talent to popularize his dreams. Mars, he believed, was home to an advanced civilization capable of a vast construction project that from Earth was 'suggestive of a spider's web seen against the grass of a spring morning'.

Lowell stubbornly refused to believe he could be wrong about the Martians. When other scientists argued that Mars' atmosphere was too thin and the temperatures too low for liquid water to exist, he simply ignored them. The shifting dark patterns on the surface he imagined to be vegetation, fed with melt water from the poles.

But in one respect, Lowell was right about Mars. He could see that it was a dry planet, full of sandy deserts. His fantasies about Martian canals were based on the notion of a planet in decline – a world that was running out of water, desperately trying to cling on to the old ways of life. We now know that there are no canals on Mars, and there never were. But Mars is a world in decline. Lowell's time frame was just 3 or 4 billion years off the mark.

BACKGROUND *Percival Lowell at his telescope, studying the canals he believed had been built on Mars by Martians.*
1 *During the long, cold nights of observation, Lowell made these sketches of the planet that was his obsession.*
2 *Map of all Mars, published by Percival Lowell in 1905, following years of study on his 24-inch telescope.*

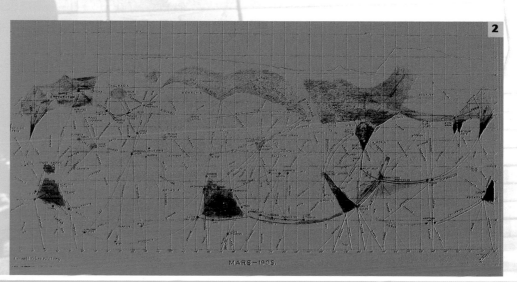

Buried in the Greenland rocks were minute black grains of carbon. Might they be the charred remains of bacteria whose rocky niche had been swallowed up by some minor geological cataclysm? Following that hunch, Mojzsis commandeered a multimillion-dollar machine called an ion probe to determine if that lump of charcoal had once been alive. The ion probe can tell the relative amounts of different isotopes of carbon in the sample. What scientists call normal carbon is an isotope called carbon-12, but carbon-13 makes up a small percentage of any lump of carbon. (A carbon-13 atom is identical in almost every way to carbon-12; in fact, you wouldn't know the difference if you swapped your normal pencil lead for one of pure carbon-13, except that it would be slightly heavier.) Living organisms, however, prefer to use carbon-12 to make up their amino acids and DNA because carbon-12 takes slightly less energy to work with than carbon-13. If a lump of carbon has more than the standard percentage of carbon-12 in it, it means that something was once alive in that lump, so it is a sort of chemical fossil of life. Sure enough, Mojzsis' hunch was right. The lumps of Greenland carbon were found to have an abnormally high content of carbon-12. These were the cremated remains of ancient microbes. But the shock came in their age: more than 3.9 billion years old.

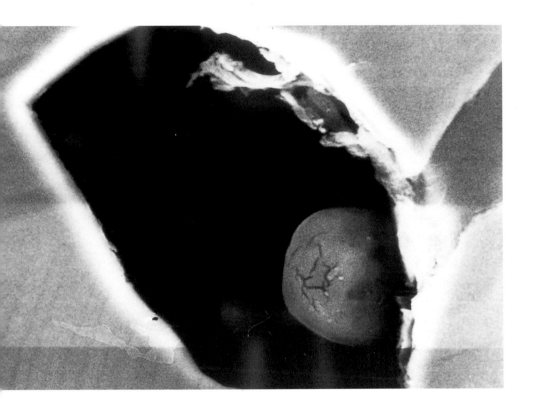

ABOVE *Stephen Mojzsis found this tiny grain of carbon (seen here magnified 1,000 times in the lower right corner of the black hole) in a 3.9 billion-year-old rock from Akilia Island in Greenland. The lump is all that is left of a colony of primitive organisms.*

The 400-million-year age difference between these and Schopf's fossils doesn't at first seem such a big deal when we're talking about billions, but it was enough to overturn long-held theories about how life on Earth began. It pushes the origins of life back into the time known as the 'late heavy bombardment' (see Chapter 1). This was the period in the history of the Solar System when death rained from the sky in the form of meteorites and asteroids thousands of times more often than at present. The planets had only recently coalesced and there was still plenty of debris floating around in what is now the empty space between the planets. Back then, the Earth was orbiting the Sun in the company of a field of giant rocks as densely populated as the asteroid belt is today. Huge meteors would strike the planet, the ground would melt and the oceans would turn to steam. This apocalyptic phase lasted until 3.8 billion years ago, when things finally started to quieten down. (This can be calculated by looking at the age of craters on the Moon: after 3.8 billion years ago the number of craters made by impacts rapidly declines.) And yet it seems that microbes were already flourishing long before the skies cleared. Life took no time at all to get a foothold here, and was born to a planet as alien from modern-day Earth as Mars.

If life started quickly on Earth, why not on early Mars too? Four billion years ago, Mars still had a thick atmosphere. It also had rainfall and flowing rivers. Even back then, Mars was probably still quite cold – more like Antarctica than the tropics – but we now know that life can exist just about anywhere there's water. The rapid rise of life on Earth raised questions of its own, however. Biologists had always thought that life needed nearly a billion years to get on its feet, in tidal pools on the edge of a warm ocean where chemicals washed in and out, slowly mixing and reacting, ultimately producing something that was primitively alive. Now the Greenland rocks reveal that life might have needed only a few tens of millions of years to develop. Maybe there is another explanation.

Seeds from heaven

In 1871 Sir William Thompson stood before a distinguished audience at the British Association for the Advancement of Science in Edinburgh and delivered a memorable presidential address. Thompson was a heavyweight scientist, soon to be given the title Lord Kelvin, but that year his speech went down badly. He suggested that meteorites brought the seeds of life from distant worlds to ours. His idea was meant to be a blow against Darwinism, which he thought was heretical. Far better that God sprayed his seed on Earth from the depths of space than that life spontaneously generated here on Earth. Kelvin's reward was widespread ridicule, especially from the Church, but his idea would persist for another century. It was called 'panspermia'.

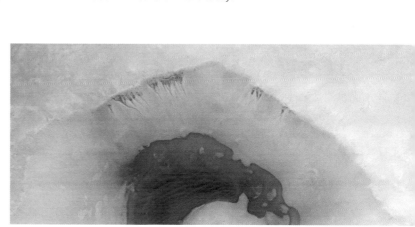

BELOW *Martian hot spring? What looks like water seepage has carved tracks down the edge of an old impact crater in this close-up picture from* Mars Global Surveyor. *Some scientists believe that springs like this could still be active on Mars today.*

The story has been told many ways in the century or so since Kelvin, but it always went something like this. A long time ago in a part of the galaxy far, far away, two worlds collided in their orbits around a bright, yellow star. One of them, perhaps it looked like the Earth, had been blooming with life. The shattered fragments from this devastating collision hurtled out across the void of galactic space, and although most were barren hunks of rock, a few contained the frozen spores of life. Millions upon millions of years pass by without event, until the path of some of these interstellar projectiles runs through a newly formed solar system. Here they once again fall under the gravitational lure of a star, this time a bluish one, our youthful Sun. In time, their newfound orbits run directly into a barren Earth, and in the terrific force of impact the dormant spores are spread far across the globe. Life can begin again. A crazy idea? Perhaps.

Kelvin based his unlikely idea on an important discovery that had been made a few years before his lecture: that certain meteorites had been found to contain organic matter – not life, but relatively complex chains of carbon and hydrogen atoms – life's building blocks. On Earth 4 billion years ago, meteorites carrying organic chemicals would have been raining down from the skies on a regular basis. Maybe that ready-made organic material helped give life a kick-start. Kelvin took this idea a step further by proposing that some of these meteorites might have actually contained dormant spores. Kelvin's question is unanswerable. But recent events have given his idea a new twist.

Half a tonne of Mars a year

One night in the late 1970s Hap McSween, a meteorite expert from University of Tennessee, sat with a colleague drinking beer out of long-necked bottles in a bar in Houston, Texas, and an amazing idea was born. McSween, then a graduate student at Harvard, had been mulling over the age of a handful of meteorites that had landed during the previous century. Meteorites are little chunks of Solar System debris that fall to Earth. They are mostly chipped-off pieces of asteroid whose orbits end up crossing the Earth's and they have almost all been found to be the same age: like the Solar System, they are 4.5 billion years old. But there was a small collection of meteorites that appeared to be only a billion years old, maybe younger. It didn't make sense. For the rocks to be younger, they had to have been melted by a geological event billions of years after the asteroids cooled to become cold lumps of rock. Everything suggested that these rocks weren't from an asteroid but from a geologically active planet. After a couple of drinks, McSween had a crazy idea: what if these meteorites were pieces of Mars?

By the time he had sobered up, the crazy idea he and his drinking partner had kicked around seemed, if anything, a little less crazy. He set about writing articles to spread the idea: a giant meteor strikes the flank of a Martian volcano with such ferocity that it catapults a handful of Martian rocks into space. After meandering around the inner Solar System for a million years or so, some of these pieces cross the Earth's path and crash-land here. Why Mars? McSween argued that of our neighbouring worlds, only Venus and Mars were big enough to have remained geologically active until a billion years ago. And since Mars is only one-tenth the mass of the Earth or Venus, it's far easier for rocks to escape the pull of its gravity and fly off into space.

It was a nice idea, but impossible to prove since we didn't have any rocks from Mars with which to compare the meteorite. No one took it very seriously. But then, one day in 1982, Don Bogard, a scientist at NASA's Johnson Space Center in Houston, was doing some routine work on one of these curiously young meteorites and found something strange. He was trying to see if the age of a small vein of green glass was the same as the rock it laced through. As his equipment analysed the sample, it detected an unusual mixture of gases coming off. Some air had been trapped in the glass for millions of years and was only now being released. But the air was not from Earth. Bogard forced himself to blink as he looked at the chemical composition of the gas. It was a dead ringer for the atmosphere that the *Viking* probes had detected several years earlier. Inside this meteorite was a pocket of Martian air!

It is now estimated that around half a tonne of Mars arrives on Earth every year – fragments blasted off the red planet millions of years ago. Almost all of it goes undetected, of course, as fine particles of dust. But there are 13 confirmed pieces of the planet Mars on Earth. One of these is the famous ALH84001, which caused a stir in 1996 because it was at first thought to contain Martian fossils. At the time of writing, the complex arguments and counter-arguments for the presence of fossilized life forms in this Martian rock continue to grind on, but the preponderance of opinion is shifting against the original claim. What is clear is that the planets we always thought of as isolated by a vast cosmic vacuum are, in fact, intermittently, but significantly, in contact. If life could escape from one planet and travel to another, how long could it exist in the harsh environment of space?

Interplanetary sneezing

In 1969 astronaut Pete Conrad went for a stroll across the Ocean of Storms. He was the third man on the Moon, and part of his historic walk included a trip over to the place where the robotic spacecraft *Surveyor 3* had been sitting since 1967. He unbolted the camera housing on the defunct ship and brought it back to the lunar module. Back on Earth, scientists made a surprising find: bacteria that had been inside the camera case since before *Surveyor* left Earth came back to life. Despite spending two and a half years in the harsh vacuum and cold of space, the bacteria had hunkered down and survived unscathed. Life, it appears, can survive in deep space.

Perhaps life started on one planet and spread to the rest. If pieces of Mars have been blasted off their home planet and landed on the Earth, we can be pretty sure there's plenty of the Earth on Mars. The early Solar System was a violent place, full of debris. Planets would be smashed on a regular basis by meteorites as big as the one that probably killed off the dinosaurs 65 million years ago. There would be chunks of Venus, Mars, Mercury and the Earth flying around all over the inner Solar System. A few pieces might even have made their way out to Europa.

Pete Conrad proved that a local version of panspermia really could work, where spores travel not from star to star, but from planet to planet within our Solar System. Primitive bacteria can survive the perils of outer space, as long as they are protected from the direct light of the Sun. Inside a camera case, inside a rock, either one affords adequate protection. Studies of bacteria that have gone into suspended animation in frozen corners of the world show that some species can survive for millions of years in this state, waiting for heat and moisture to bring them back to life.

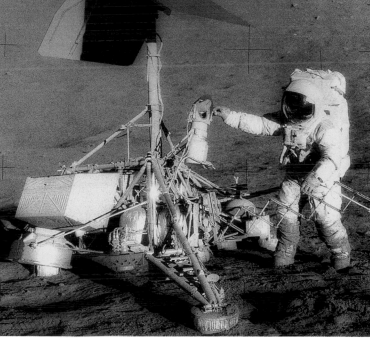

ABOVE Apollo 12 *astronaut Pete Conrad grapples with defunct probe* Surveyor 3's *camera and unwittingly discovers Earth bacteria that had survived inside the case for two and a half years.*

A new belief is starting to take hold among exobiologists: if life started on Earth, chances are that it started on Mars too. If life didn't begin on its own on both planets, it would just be a matter of time before the Earth sneezed life on to Mars, or vice versa. Life on Earth may have Martian ancestry. Of course, to find out for sure we have to go to Mars and hunt for fossils.

The greatest discovery

There are some faint hopes that life could still exist on Mars, that it might have retreated from the surface as the planet cooled to a safer refuge under ground. Deep down, where the planet's internal heat is enough to keep water liquid, perhaps life is clinging on, just as bacterial life has now been found on Earth living in rocks kilometres below the surface. But if life ever started on Mars, there's a good chance of finding fossils of it strewn about the surface. Mars is not as active as the Earth. You don't have to go to a remote Martian Greenland to find rocks 4 billion years old – half of Mars is that ancient. We even know roughly what to look for: Imre and Roseli Friedmann's algae inside Antarctic rocks leave fossils that are easy to recognize by their mottled colours.

A quarter of a century on from the disappointment of *Viking*, NASA is gearing up for another onslaught on Mars. *Pathfinder* and the orbiting *Mars Global Surveyor* opened the way for a decade of missions that will look for life on Mars. Some time in the early years of the third millennium, one or two robotic rovers will make an extensive search of the red planet, looking for rocks that just might have fossils of primitive Martian life sealed inside them. If they find

ABOVE Mars Pathfinder*'s panoramic view, taken in 1997, of the robotic* Sojourner *vehicle by the large rock* Yogi *on the surface of Mars.*

something, their roving is over. They will stop and wait for another craft from Earth to land nearby, collect the sample and bring it back home. If they find a rock that proves there was once life on Mars, it will be, without any doubt, the greatest scientific discovery ever made.

But Mars is not our only hope. In November 2004 the European Space Agency's *Huygens* probe will parachute down on to the surface of Saturn's moon Titan. Although Titan is certainly too cold for life, some answers to the mystery of life's beginnings might be hiding there. Over the aeons, countless meteorites are bound to have hit this icy moon, and when a meteorite strikes, the ice melts. Suddenly an impact crater is a small lake, hot enough for water to remain liquid there for 1,000 years. It's not much time, but perhaps enough for the first chemical steps towards life to take place. And what if, countless years later, *Huygens* landed on top of that now refrozen lake? Beneath it could lie the secret of our origins.

And what of Jupiter's moon Europa? The possibility of life in an ocean below its ice covering is so compelling that this once insignificant moon is now one of NASA's top priorities. 'If we could go tomorrow, we would,' said one mission planner. A Europa orbiter armed with a radar to see through the ice crust and look for positive proof of a subterranean sea will certainly fly in the first few years of the 21st century. If there is an ocean on Europa, we might one day mount our most ambitious space mission to date to look for life in it. In 2015, perhaps, a spacecraft could touch down on the frozen crust. As the brittle ice crackles beneath its feet, the probe will turn on a heating device and begin to melt its way down to the ocean below. Once through the crust, a small tethered submarine will descend into an alien sea and start its search. Only then could we truly say we have dipped our toes into the cosmic ocean.

Little green men

The search for signs of life beyond the Earth is full of false alarms. When the flickering stars called pulsars were discovered in 1967, astronomer Jocelyn Bell named them all with the prefix 'LGM', standing for 'little green men'. She was only half joking – the possibility that pulsars were alien beacons was one she had to carefully eliminate before announcing her findings. If only all her colleagues in science had been so careful…

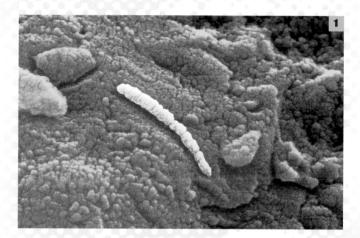

1871 German part-time geologist Otto Hahn claimed to find peculiar patterns on almost every meteorite he laid his hands on. Hahn announced his discovery of fossilized ferns, corals and sponges. But he was soon made the object of scientific scorn – the patterns were nothing more than mineral formations.

1899 Electrical pioneer Nikola Tesla detected what he thought were signals from Mars at his laboratory in Colorado. It seems he might have been picking up Guglielmo Marconi's test transmissions of the wireless across the English Channel.

1921 It was Marconi's turn next, and the *New York Times* announced that the eminent pioneer of radio had picked up what might be coded signals from aliens while aboard his research yacht. The source was probably natural.

1930 'Life in Meteors' announced by the *New York Times*. Charles Lipman, a bacteriologist at the University of California claimed to have grown living cells from meteorites in his laboratory. Lipman was no fool and realized that the meteorites would have become contaminated after they landed on Earth, so he had sterilized them. Apparently, he didn't do it well enough. When others repeated his work, the cultured bacteria turned out to be all too familiar and terrestrial.

1961 Bartholemew Nagy and George Claus found 'organized elements' of organic matter in meteorites. When other scientists examined them, they turned out to be either down-to-earth pollen grains that had blown in through the window, or strangely shaped mineral grains. The spirits of Lipman and Hahn return.

1996 Scientists at NASA's Johnson Space Center announced the tentative discovery of fossilized microbes in meteorite ALH84001, known to be from Mars. Small, worm-like objects could be seen on parts of the rock when it was placed under a high-powered electron microscope. Other scientists contended that the 'worms' were either contamination, or an artefact of the laborious preparation that the rock had to undergo before it could be studied under the electron microscope.

1   *Electron-microscope image of the Mars meteorite ALH84001. This 'worm' is most likely a terrestrial contamination.*
2   *Marconi beams the news of alien signals around the world.*

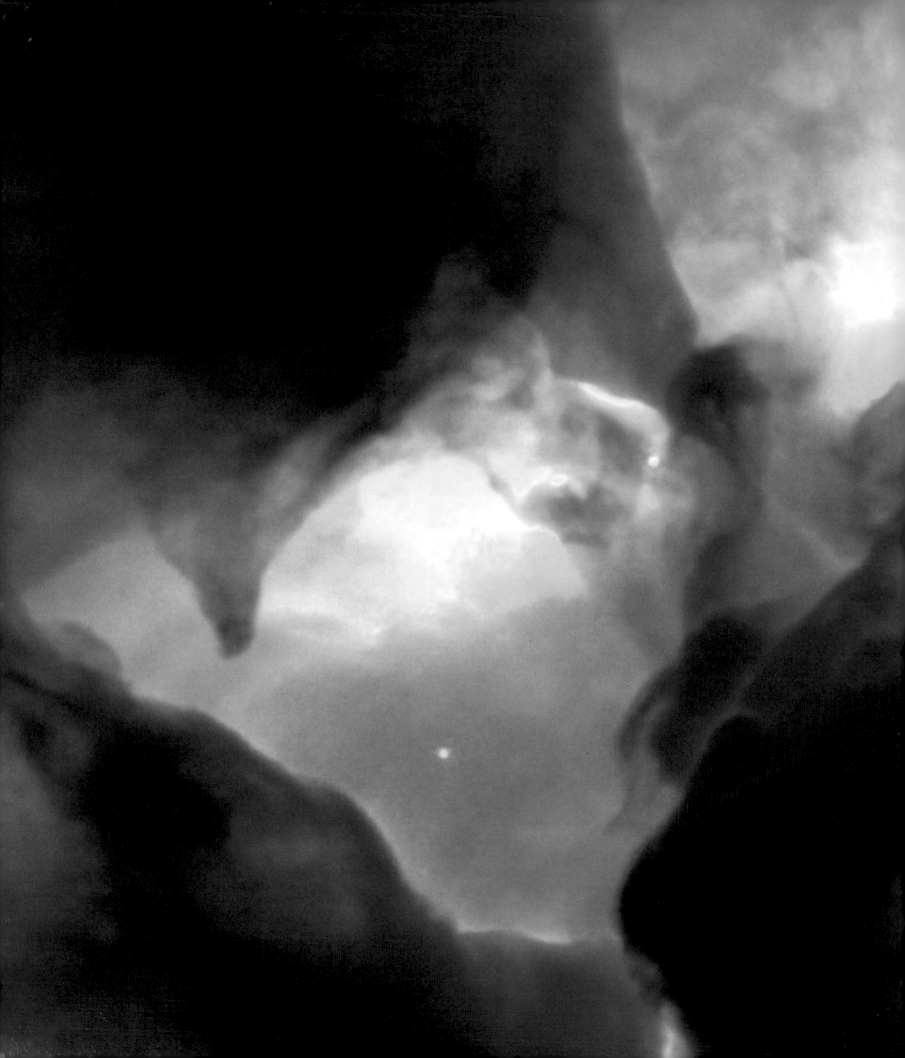

beyond the sun

MORE THAN 5 BILLION kilometres from Earth, the imaging platform of an old spacecraft creaks and groans back into life. It has just been awakened by a message, the command sequence for its first mission in nearly a decade. *Voyager 1's* camera hasn't seen use since it hurtled by the rings of Saturn in 1981. But that encounter flung the craft up and out of the plane

inhabited by the planets and now, speeding away at 60,000 kilometres per hour, it has an enviable bird's-eye view of the Solar System. Planetary scientist Carl Sagan has spent years persuading NASA bosses that a portrait of the Sun and its planets is worth their time and effort. It is not really science; it is a piece of romance, the chance to snap a picture of the Earth and its planetary siblings as they would appear to a star-traveller approaching our corner of the galaxy. Now that the *Voyager* team is about to disband, Sagan has finally got his way. On this St Valentine's Day 1990, *Voyager's* lens has started its hunt for those distant specks and will soon be radioing back to Earth a picture with a caption that reads: 'The Solar System. Age: 4.6 billion years.'

Over the course of three months following that Valentine's Day snapshot, *Voyager 1* slowly sent back images of a faint Sun and six of its planets: Earth, Venus, Jupiter, Saturn, Uranus and Neptune. Mercury was lost in the glare of the Sun, and both Mars and Pluto were too faint to register anything at all through *Voyager's* straining eye. To the imaging team this mosaic of 60 pictures, cutting a snaking path across the Solar System, came to be known as 'the family portrait'. In the photographic labs at the Jet Propulsion Laboratory (JPL), in Pasadena, California, they tried to put a scale view of the mosaic together, but at their first attempt the planets were too small to be seen even on

PAGES 214–15 *The Lagoon Nebula in the constellation of Sagittarius – a cloud that is both the nursery and the graveyard of stars, and possibly other planets.*

the highest quality paper. In the end they took the images of the smallest planets – Earth and Venus – and scaled them down until they could just be recognized as dots of light. Working at that minimum level of detail, when the mosaic of images was put together to scale it stretched across a piece of paper 6 metres long before reaching Neptune. Such is the size of our Solar System.

The family portrait is not, in truth, a fine picture. If you look at the Earth, you won't see clouds or continents or even oceans, just a pale, slightly blue dot. Uranus and Neptune aren't even dots; they are smeared by the spacecraft's motion. But that is to miss the point. Here is a picture frozen in time: the Sun's worlds as they appeared for a few seconds on 14 February 1990. Had the technology been available to take those pictures 4.5 billion years previously, we

ABOVE *Lost in the stars: the Sun is just one star amongst hundreds of billions in our galaxy. This image shows just a few of our nearest neighbours.*

would have seen an entirely different group of worlds. We would have seen the Earth and Venus surrounded by steam, Saturn without any rings, and Mars with active volcanoes and flowing rivers.

  The Solar System posed for its first picture when it was well into a ripe middle age. Only human myopia or arrogance could lead us to believe that things will always look the way they do now. Nothing escapes the ravages of time. In this last chapter we look to the future. How will that picture change over the coming aeons; what worlds will one day circle our ageing Sun?

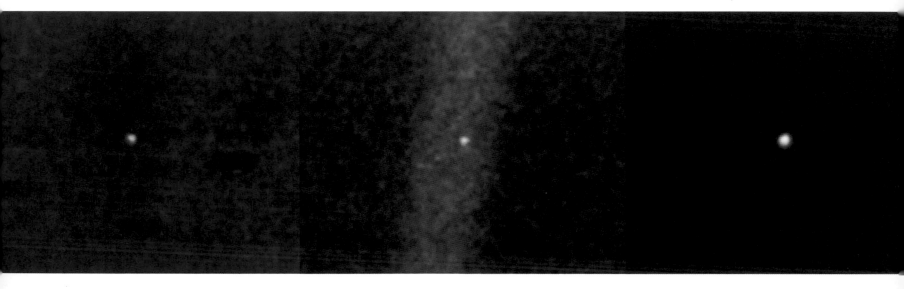

One billion years AD

Flying back to those dots that lie close to the Sun a billion years from now, we might well find ourselves approaching a planet shrouded in a thick veil of gas. Our instruments would detect carbon dioxide and perhaps some trace of water in the air. The temperatures below the clouds would be hundreds of degrees Centigrade. The pressures would be unbearable for life. No, this is not Venus. Rather, Venus has a new sister planet: this is the Earth in a billion years' time.

  This possible future for the Earth is the work of the Sun. The object we think of as stable from day to day and year to year is not, in fact, immune to the effects of age. In about 1 billion years from now the Sun will be 10 per cent brighter. According to most calculations, this will spell disaster for the Earth as we know it, sending it the way Venus went several billion years ago – into the grip of a runaway greenhouse effect. Water from our oceans will begin to evaporate as the temperature creeps up, and the more steam there is in the air, the better this greenhouse gas will trap heat. And so, as the temperature spirals upwards, the Earth's oceans end up in the sky and our world turns from cradle into cauldron.

This nightmare vision is only one of the possible fates awaiting the Earth. Other events might take precedence. There is, of course, the ever-present danger of a large meteorite wiping out life on our island in space. But there is a more predictable event in our future. One day, the molten insides of our planet will cool to a critical temperature, at which point they will suddenly turn solid. No geologist can say precisely when this will happen; it could be a billion years from now, or it could be much sooner. And although this won't mean the end of volcanic activity or the end of plate tectonics – the Earth's mantle and core will still be very hot – it will instantly wipe out our planet's magnetic field. As explained in Chapter 5, the spinning liquid metal at the

BELOW *The family portrait – Voyager 1 took these snaps of Mars, Earth, Jupiter, Saturn, Uranus and Neptune when it was 5 billion kilometres from Earth – the pale blue planet second from the left.*

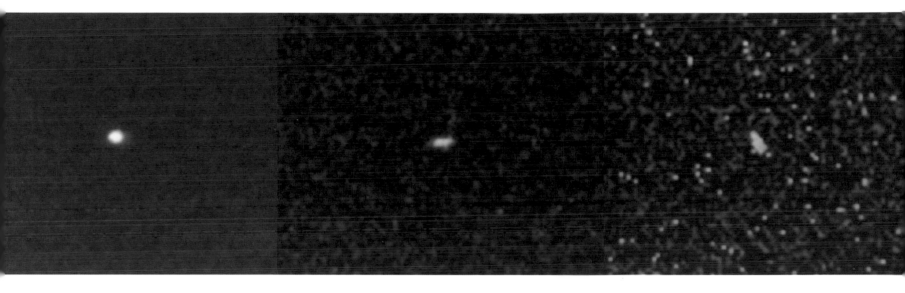

Earth's core generates the magnetosphere, our protection from the ravages of solar wind. Without this invisible force field, our atmosphere will begin to be stripped away into space by the barrage of particles streaming out of the Sun. In this possible future the Earth risks becoming a dry, almost airless world, more like Mars than Venus.

Brave new worlds

Whichever way we look at it, the Earth seems doomed. But it is not the only world to change, and it is not all bad news. As the Earth loses its sparkle, Mars might become a diamond in the rough. In a billion years, the extra heat pouring out of the Sun could unlock water frozen below the surface of the red planet. Rivers could flow again down the valleys that *Viking* and *Mariner* found, skies could change from pink and dust-laden to blue, full of water vapour and carbon dioxide released from the polar caps. Perhaps the human race, if it still existed, could skip from the blue dot to the red one as the Sun changed its ways. And if they were to look up at Mars' night sky, our descendants would not find the planets we now know.

Saturn's sepia-coloured clouds will still course around its circumference in subtle bands, but the bright rings might be missing. Some scientists believe that in about 100 million years from now, the billions of house-sized chunks of ice in the rings will have begun gently pulling together. As they group, these clusters will exert a gravitational influence on the bodies around them. Before long, the neat orbiting rings of debris will be fatally disrupted. Some of the smaller lumps of ice will plunge headlong into Saturn itself to become part of its gaseous cloud tops. Others will either drift away or be spun out into the darkness as a result of the gravitational tug of war that will ensue. What was the most beautiful sight in the skies will become a denuded brown ball.

By that time, Neptune might well have usurped Saturn's role as the most eye-catching planet in the Solar System. Triton, Neptune's beguiling moon, is unlike any other major satellite in the Solar System, orbiting against the spin of its parent planet. The result is that the tidal pull of Neptune is slowly but surely pulling Triton towards it. But Triton's fate is not to be swallowed up by the ice giant, rather to be torn apart by it. In Chapter 4 we described the invisible line in space beyond which no moon should cross – the Roche limit is the point at which the gravitational forces of a nearby planet on the inside and the outside edge of a moon are sufficiently different to pull the moon apart. This is Triton's destiny – to be shattered by the gravitational muscle of the looming blue globe. But Triton is a massive moon, far bigger than the one that broke apart to form Saturn's rings. In the far reaches of the Solar System, this pale blue gem of a planet could one day be set inside a dimly glistening necklace that dwarfs the rings we see today around Saturn.

In the realm of a giant

From the perspective of a race that has gazed up at the same sky for its entire communal memory, these changes seem dramatic, even cataclysmic. They are not the final chapter, though, by any means. The ultimate fate of the Solar System lies firmly in the hands of the body that gave birth to the planets. The Sun cannot last for ever; that much we have accepted for quite some time. But how, and when, will the Sun play its final hand?

Back in 1868, bent double over his spectroscope on a platform high above the church of St Ignatius in Rome, Father Angelo Secchi was peering at the night sky. He knew that not all stars emitted light of exactly the same colour. Some, like Vega, had a distinctly bluish tint; others, like our own Sun, were stronger in the yellowish portion of the spectrum. Secchi had, by then, published a catalogue of over 300 stars listed by their colour. But in that year he discovered a new type of star that was tinged fiery red. After checking the exact shade of these red stars against the colour of the flame on a simple Bunsen burner, he realized they were both producing the same chemical element. He named his celestial discovery a 'carbon star'.

Eventually, astronomers came to realize that the array of different coloured stars Secchi had categorized weren't all fundamentally different types of stars – some represented different stages in the evolution of stars. Although human beings could never hope to witness the birth and death of one star, the night sky provided an album of snaps spanning the entire life cycle of stars. As the 20th century advanced, so too did our understanding of the processes that drive the nuclear engine at the heart of stars. Soon the theoreticians could plot the life story of any star, no matter what its size. In his carbon star, they realized, Secchi had glimpsed the future of the Sun.

As we saw in Chapter 5, a star forms from a collapsing cloud of mostly hydrogen gas that contracts under its own weight until the pressures and temperatures at its core initiate a self-perpetuating nuclear reaction. For billions of years, deep in the dense core, hydrogen atoms fuse together to form heavier atoms of helium. During this stage in the star's life, it can be identified by its yellowish tint. But one day the star's supply of hydrogen starts to run short and the core fills up with a mass of superheated, super-dense helium. Now the helium core begins to contract, and as it does so, it heats up to the point where helium atoms can be forced to fuse together. This is the final chapter in the evolution of a star the size of the Sun. It is now a body that burns helium at its core to produce carbon. It is many times hotter and brighter than the Sun is today, and it has expanded to 100 times its present size. The star is now a red giant.

In the dying days of the Solar System, beginning about 6 or 7 billion years from today, the Sun will lurch into its final decline, and it has a fiery end in store for the planets. The surface of the scorching red star will balloon out to reach beyond the orbit of Mercury. But well before it expands that far, Mercury will be little more than a memory. The world famed by the Ancients for its fleet-footed passage across the night sky will have long since been slowed by the encroaching atmosphere of the ageing star. All too quickly the drag on the tiny planet will have caused it to fall into the Sun. The Earth, Mars and Venus will probably escape wholesale consumption, but their prospects are dim. The increasingly violent solar wind – tens or hundreds of times more powerful than today's – will have denuded first Venus, then the Earth of their thick shrouds of gas. Their rocky surfaces, blistering in the intense heat, will be laid bare and lifeless. Soon afterwards, Mars, too, will suffer the same fate.

A fleeting thaw

Beyond the asteroid belt, the growth of the Sun from a bright star to a red disc might not be such bad news. The icy moons of Jupiter and Saturn will be warmer than they have ever been since the first days of the Solar System. Although the icy surface on Jupiter's Europa will probably remain frozen solid, further out at Saturn's moon Titan, events might take an interesting turn.

Two scientists from Arizona, Ralph Lorenz and Jonathan Lunine, along with Chris McKay from NASA Ames in California, have proposed that Titan could enjoy a 500-million-year thaw, during which it might even become a habitable world. Because Titan is blessed with a thick atmosphere, it will have the ability to trap the heat of the old red Sun. Today the temperature at the surface of this moon hovers around a chilling minus 200 degrees Centigrade. Six or 7 billion years from now, with the help of the greenhouse gas methane and

---

Probes still in space (continued)

USSR
1959 Luna 1
1961 Venera 1
1962 Mars 1
1963 Luna 4 – in orbit around the
   Earth/Moon system
1964 Zond 1 and 2
1965 Luna 6
   Zond 3
   Venera 2
1966 Luna 10, 11 and 12 – Moon orbit
1968 Luna 13 – Moon orbit
1971 Luna 19 – Moon orbit
   Mars 2 – Mars orbit
   Mars 3 – Mars orbit
1973 Mars 4
   Mars 5 – Mars orbit
   Mars 7
1974 Luna 22 – Moon orbit
1975 Venera 9 and 10 – Venus orbit
1978 Venera 11 and 12 flight platform
1981 Venera 13 and 14 – Venus orbit
1983 Venera 15 and 16 – Venus orbit
1984 Vega 1 and 2
1988 Phobos 1
   Phobos 2 – Mars orbit

JAPAN
1985 Sakigake
   Susei
1998 Nozomi (Mars orbit 1999)

EUROPE
1985 Giotto

Of course, the four spacecraft that have left our Solar System should not be forgotten:

USA
1972 Pioneer 10
1973 Pioneer 11
1977 Voyagers 1 and 2

The life and death of stars

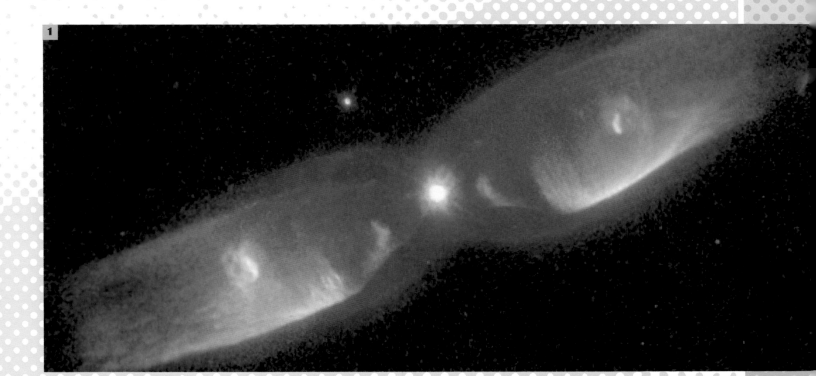

Stars are born from clouds of gas which contract into dense balls. The lifespans of stars depend on their size: the bigger they are, the shorter they live. In the initial stages of collapse, as material starts its gravitational plunge towards the centre of the sphere, the temperature rises and it starts to glow with a steadily increasing intensity. Eventually, the heat of contraction reaches a critical temperature – in the case of our Sun, it was about 15 million degrees Centigrade – and nuclear reactions kick in as hydrogen is converted to helium. The immediate effect of this nuclear generator is to create a gust of wind which blows away the gases still drifting in towards the star. The second effect, paradoxically, is to stop the star overheating: the increase in core pressure prevents the inward contraction of gases, thereby halting the temperature rise. All this happens in a remarkably short space of time, around 1 per cent of the life cycle of a star, and it quickly enters a long period of stability. Eventually, typically within a few billion years, the hydrogen at the core of the star is completely converted into helium, the ball of gas once again begins to contract and the temperature surges. Now the hydrogen in the once-cooler outer layers begins to burn and, in the core, helium is converted to new elements such as carbon. This new burst of nuclear fusion causes the star to expand dramatically. Our Sun will exist as a red giant for about a billion years. Gradually, as the star continues to heat up and new types of nuclear reactions explosively kick in, the star sheds its outer layers back into space, leaving behind a glowing ball of dense matter known as a white dwarf. All nuclear reactions have now ceased and the dwarf slowly leaks heat out into space like a burnt cinder until its light goes out forever.

1   Over 2,100 light-years away a star goes out in blaze of glory. The Hubble Space Telescope captured this beautiful image in August 1997. One day, our Sun could end up looking like this.

a giant red Sun pumping out many times more heat than it does now, the surface temperature could creep up higher than minus 100 degrees. This might not sound especially inviting, but it means a lot to this frigid world. Titan's present-day surface is probably a frozen mixture of water and ammonia; warm that by 100 degrees and you have a pungent ocean of diluted cleaning fluid.

This ocean might not be much like the Earth's original primordial soup, but being rich in organic chemicals, it could nevertheless take a short walk down the path towards life. For half a billion years a second experiment in the origin of life might take place. It is highly unlikely, however, that any human will be around to witness it.

After the Sun has been a red giant for around half a billion years, it will undergo the final apocalyptic event of its life. In a last desperate gasp of activity, the Sun will shed the outer layers of its gas out into space, treating the dying planets to one last great roar of solar wind thousands of times more intense than today's. No scientist dares predict how the planets will react to a quarter of the Sun's mass being blown past them. Suffice to say that what is left of the Sun afterwards – a dense white-hot core where nuclear reactions are now a thing of the past – will never again be enough to warm them, to stir their clouds or melt their icy surfaces. On this day, when the Solar System is around 12 billion years old, it will be pronounced dead in the great log-book of the galaxy.

The next frontier

On a Thursday morning in February 1600 another death was recorded. Giordano Bruno had just been burned at the stake in Rome for the heresy of suggesting that the Sun was not the only star to have planets. He had spent eight years in a dungeon for this and other beliefs that were considered utterly outrageous by the Church. For Bruno, each star in the night sky was another Sun, each one had its own Earth, and each Earth had its own Jesus. Bruno's claims came a full decade before Galileo dared to make his comparatively tame assertion that the Earth revolved around the Sun. By a long way, Bruno was ahead of his time.

Could the cosmic counterparts of the Earth and Jupiter be dotted freely across the heaven of stars? Might distant solar systems survive long after ours is dead? Bruno's idea remained nothing but speculation and fantasy for centuries. Planets around other stars – even if they existed – would be just too small to see, even with the best telescopes we have today. But as we saw in Chapter 1, the first sign of an alien solar system appeared in 1984, when two astronomers snapped an infrared picture of a cloud of dust swirling around the star Beta Pictoris.

Beta Pictoris showed no sign of planets, but something about the infrared picture was very curious. Scientists think that parts of this cloud of dust might one day coalesce to form a family of planets like our own. In fact, there were signs that it was already happening. The

TOP *The expanding outer atmosphere of the red giant, Betelgeuse – 300 times the diameter of our Sun. One day, our Sun too will bloom as it enters the final stages of its life.*

ABOVE *The burning of Giordano Bruno (1548–1600). The Italian philosopher died for suggesting that around other stars there might be other Earths with their own versions of Jesus.*

dust disc of Beta Pictoris was seen from the side, so it was not easy to tell whether sections of it had already condensed into planets. But there was a hint that this was well under way. None of the dust in the disc was very hot, meaning that none of it could be very close to the star. The very inner portion of the dust ring must have been swept clear: there could already be a family of young planets circling close to this star, eating up the dust. It was an enticing idea, but it fell short of definite proof of the existence of planets beyond the realm of the Sun. Since that first sighting, other clouds of dust have been seen around other stars. Some of them face the Earth directly and, in a few cases, have allowed infrared astronomers to see a gap cleared close in to a star.

The definite proof of other solar systems finally came on 3 July 1995. That day will be remembered for the most important event in planetary science since the Copernican revolution that put the Sun at the centre of the Solar System. At an observatory in the French Alps two Swiss astronomers, Michel Mayor and Didier Queloz, had spent months peering at a star that was behaving very strangely. The star, 51 Pegasi, in the constellation Pegasus, had been wobbling. It was not a wobble that was big enough for them to see directly, but it had a tiny effect on the colour of light from the star. Just as the siren on an ambulance changes pitch as it moves past you, so astronomers see a subtle colour change when a star shifts between moving ever so slightly towards the Earth and ever so slightly away from it. The wobble that Mayor and Queloz thought they had been picking up was really minuscule: the star was juddering at only 10 or 20 metres per second. But there seemed to be a pattern to the wobble – the behaviour was repeating every four days or so.

Mayor and Queloz had a hunch that a planet orbiting around the star could be causing the movement. As the planet went round, its gravitational force would yank the star after it. The star would certainly be many times more massive than the planet, so the resulting movement of the star would be almost imperceptible, but it could be just enough to show up as these small colour changes they were seeing. If a planet was indeed the cause, Mayor and Queloz thought they ought to be able to predict exactly what colour the star should be at a given time that night of 3 July, based on the observations they had made months earlier.

Late that night the champagne started to flow and the next day the pair of astronomers had written down the vital statistics of a planet. It is a body about half the mass of Jupiter, its year lasts 4.2 Earth days, it flies through space four times faster than our planet and hovers closer to its star than Mercury does to our Sun. The observations had been a complete success. Mayor and Queloz had discovered the first world outside our Solar System.

Wandering giants

At the time of writing, at least 17 other planet-like objects have been detected in orbit around other stars. All but two of them, including the planet around 51 Pegasi, are like nothing in our Solar System. Around the star 47 Ursae Majoris there is a planet about two and a half times the mass of Jupiter, orbiting at a distance further out than Mars orbits our Sun, while at HD210277 there is a planet one and a third times the mass of Jupiter, orbiting just further out than the Earth does from the Sun, and having a year that lasts 437 days. But the rest are in orbits so tight and are so massive that they defy all our understanding of how planets form. Perhaps the most alien is the giant planet, nearly four times the mass of Jupiter, which orbits the star Tau Bootis in just 3 days and 7 hours. It is so large and so close to the heat of its Sun that scientists think its clouds must be made of vaporized rock.

Until 1995 it had always been thought that giant planets could form only far away from their parent stars. Close to the star, it was argued, any gas would be too hot to coalesce into a solid body. As we saw in Chapter 1, from our parochial experience in our own Solar System, the gas giants form only where the temperatures are cool enough for water to freeze solid. But there might be an explanation for finding giant planets close to their stars if we're prepared to accept that where they are now is not necessarily where they formed.

It is possible, according to the latest theories, for giant planets to form far away from their host star and then to spiral inward, feeling the frictional drag of the disc of dust and gas around them. Some proponents of this idea suggest that Jupiter was the last in a long line of great planets to form in our system but the first to survive falling into the Sun. Perhaps the giant planets we can now 'see' out in deep space have somehow managed to stop themselves right at the brink of such a fiery death.

It is shocking how different the new planets are from the old familiar nine. In the growing family of solar systems we now know, ours looks decidedly like the odd one out. We have no giant Jupiters scooting around inside the orbit of Mercury, no planets with revolutions best measured in hours rather than years, no clouds made of vaporized rock. But that's not to say there aren't distant solar systems like our own. It's just that we are barely able to detect them yet. Our Sun is being wobbled by the planets at a speed of about 12.5 metres per second. But since it is Jupiter that's mostly responsible, the wobble takes one Jovian year, or nearly 12 of ours, to finish one cycle. Astronomers from a distant solar system would have to watch our Sun for 12 years before they could say that there was a giant planet in orbit around it. Clearly, stars with wobbles lasting days are much easier to spot. And why are the new planets so massive? Again, that's because of the way we are looking, not because of the way things necessarily are around the galaxy. The Earth makes our Sun wobble at 9 centimetres a second, which is about 30 times less than the smallest wobble our telescopes are able to detect. We're finding big planets because they're the only ones that pull their parent stars around fast enough for us to notice.

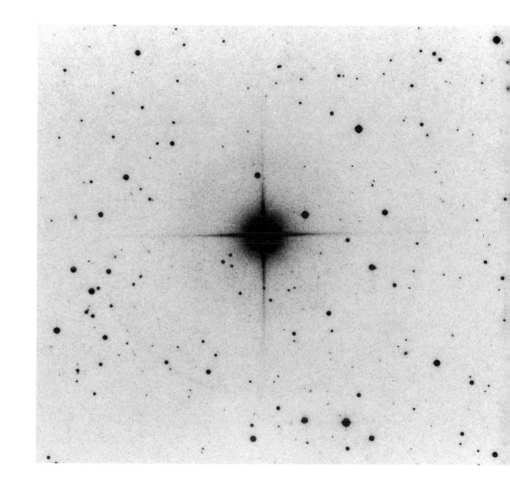

ABOVE *The star 51 Pegasi was a bright beacon of hope to astronomers in search of other planets. In 1995 it became the first star to be confirmed to have a planetary companion.*

Twinkle, twinkle, little star

In just a few short years, planets beyond the Sun have become a commonplace idea. One recent survey scanned 107 stars like our Sun and found planets around six of them — who knows, there could be billions of worlds out there. It seems tantalizingly possible that there exist thousands of planets like the Earth, that there are many places in the galaxy where the conditions could be suitable for life. The challenge of finding them has suddenly become one of the top priorities in the field of astronomy.

There's no way to find Earth-sized worlds by looking for wobbling stars. But there are other methods, perhaps more obvious, even if they are technically harder. If a distant planet passes between us and any star we are looking at, it will cut off a small fraction of the light we see from that star. The star will wink at us ever so slightly, changing its brightness by a tiny fraction of a per cent. The effect is so small that the Earth's atmosphere would completely obscure it; a gentle gust of wind over the telescope would cause a much bigger twinkle. From space the chances would be much better: all you have to hope for is the perfect alignment of a planetary orbit with the line of sight from Earth. In the year 2001 the French space probe *Corot* will head into orbit to scan the heavens for signs of twinkling stars. It won't see the planets themselves, but it might spot their silhouettes.

Even before that, it might be possible to catch the fleeting traces of new Earths from the ground. Stanton Peale, a planetary scientist in Santa Barbara, California, has proposed that a bizarre effect predicted by Albert Einstein's theory of relativity, called 'gravitational lensing', could come to the rescue. As a planet the size of Earth passes between its star and our lookout post light-years away, it momentarily bends the light from the star, acting as a tiny lens. For just a fraction of a second, the star will flash two or three times brighter than normal — bright enough to be seen even from ground-based telescopes. As yet, no telescope has attempted to survey the sky looking for these bright flickers, but the task should be under way by the turn of the millennium. Within a few years at most, the discovery of the first new Earth-like planet could be making headlines the world over.

What's in a picture?

When Galileo saw those four bright dots circling around Jupiter in 1610, he proved to the world that seeing really is believing. All of Copernicus' painstaking calculations were good in theory, but nobody was truly convinced that the Earth was not the centre of the Universe until that Florentine night of revelation. Four moons in an orbital dance around a another world. How many words is a picture worth? How many plots of orbital wobbles of stars would scientists readily give up for one clean shot of an alien planet?

The Hubble Space Telescope is a marvellous invention — a pinnacle of 20th-century technology. From its vantage point 600 kilometres above the Earth, it is able to make out mountains on Mars just 30 kilometres across. But to take a photograph of a mountain on a

ABOVE *In 1997 the Hubble Space Telescope took this image of a newly formed binary star and saw something unusual. At the end of a bright streak of glowing gas and dust extending from the star is a faint object, too dim to be an ordinary star. Could this be a giant planet, several times bigger than Jupiter, that has been thrown out of this young solar system?*

planet around another star, a telescope the size of the USA would have to be launched into space. That's not going to happen for a while, but a fuzzy picture of one of the new giant planets might not be so far off.

Planets are hard to see because they don't produce their own light, they only reflect it from their star. Trying to find a planet next to a star is like looking for a glowing ember in a roaring furnace. When you look in infrared light, which planets emit because of their own internal heat, the situation is not much better. In our own Solar System, the Sun is a billion times brighter than the planets in visible light, and 10 million times more so in infrared. Neither option is what you would call a good bet, but looking for a planet in infrared light is clearly the way to go.

By the year 2000, two telescopes on Earth — one in Arizona and one in Hawaii — will have a fair chance of taking an infrared picture of one of the giant planets that the wobbling-star gazers are so sure exist. Both telescopes will use a trick to see very faint, small objects; they actually consist of two mirrors spaced a short distance apart and rely on the fact that light acts like a wave. Both mirrors are precisely focused on a star and, by incredibly accurate positioning, the peaks of the light waves in the star's image from one mirror are made to

ABOVE *Building solar systems. The discovery of giant planets around other stars has made theoreticians rethink the way planets might form. In this model from scientists in Santa Cruz, California, a Jupiter-sized planet has carved out a groove in a disc of gas and dust around a young star. The planet's presence is sending ripples through the nebula, which might lead to the formation of more worlds.*

combine with the troughs from the same image in the other mirror. In the resulting composite picture the star disappears – the image from one mirror cancels out the other. But the cancellation only works for the object right in the centre of the image. If there is a planet just next to the star, it should suddenly appear from the ghostly darkness.

To take a picture of an Earth-sized planet would require combining mirrors so that they cancel out starlight even more effectively. To see such a faintly glowing object would take an array of floating mirrors spanning more than 60 metres in space, away from all the distractions in the Earth's atmosphere. Such a bold mission, called the *Terrestrial Planet Finder,* is under serious consideration by NASA and European nations. Some time between 2010 and 2020, *Planet Finder* could be making its way to an orbit around the Sun 50 million kilometres from Earth, or perhaps even venturing out to Jupiter to begin its patient watch even further from the heat of our own Sun.

To see or not to see

When it comes, the picture could trigger a million possible questions, but only one of them is inescapable. Is there life on this new Earth? As discussed in Chapter 7, if we accept liquid water as the first necessary ingredient for life, a planet that's likely to be living can't be too far from its star nor too close to it. It must also be big enough for its gravity to be sufficient to hang on to an atmosphere. The picture will help us answer these questions, but they are only the beginning.

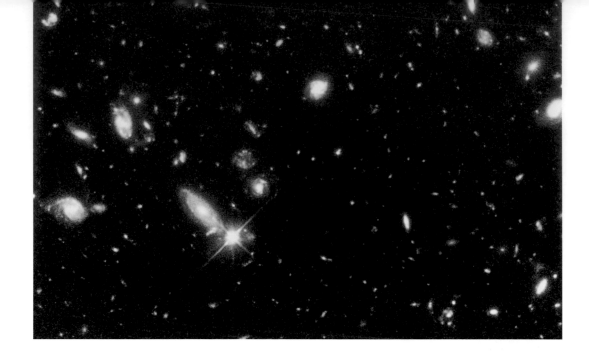

In 1993 the *Galileo* spacecraft flew by the Earth and performed a rather churlish experiment – to look for life on our own planet. It didn't fly close enough to reveal circular patterns of irrigated fields in the desert or the rigid criss-crossed maze of city grid systems which are now familiar from satellite images. *Galileo* saw the azure of the oceans and the brownish mass of the continents, but nothing in these admittedly stunning pictures gives away the existence of life. Instead, it was a few bumps in a rather dull squiggly line that clinched the experiment. When the spectrometer on board *Galileo* analysed the atmosphere of the whole planet, it detected signs of oxygen in the red portion of the spectrum. Then, looking in the infrared, it found traces of methane.

These two gases in our atmosphere are like a pair of glaring beacons that would alert any intelligent alien that the Earth has to be alive. Oxygen is a violently reactive gas that would rapidly disappear from the air unless algae and plants were continually re-supplying it in vast quantities. To find methane alongside oxygen only reinforces the case for life, since oxygen destroys methane instantly on contact. Only a constant source of methane from bacteria and animals on the surface could explain its presence.

Galileo is a dry run for the tests we will perform on the distant planets as soon as we capture their fuzzy images. There will be no hope of spotting any alien cities on those indistinct discs. But just as Father Secchi first picked apart the light of distant stars with his prism in the roof of a church, so astronomers will one day command their telescopes floating out in space to dissect the glow of faraway planets. They will look for the traces of gases that could bring mankind the most stunning news it has ever received.

The story of planetary discovery began in 1781 with William Herschel's sighting of the body that would soon be called Uranus. With that, Herschel doubled the size of the Solar System. In 1930 Clyde Tombaugh saw the fleeting presence of Pluto on a pair of photographic plates and pushed the boundary further. Tombaugh set the seal on the nine planets we think of as our local Universe, but the boundary is a fiction. We have already discovered new planets far more distant. Perhaps 100 years will pass between Tombaugh's moment and the discovery of a planet that could truly be called the Earth's twin. Then the entire galaxy will become just part of the neighbourhood.

Chronology of lunar and planetary exploration

1957
USSR Sputnik 1 (4 Oct 1957)
Earth Orbiter

USSR Sputnik 2 (3 Nov 1957)
Earth Orbiter

1958
USA Explorer 1 (1 Feb 1958)
Earth Orbiter

USA Vanguard 1 (17 Mar 1958)
Earth Orbiter

USA Pioneer 0 (17 Aug 1958)
Attempted Lunar Orbit (Launch Failure)

USSR Luna 1958A (23 Sep 1958)
Attempted Lunar Impact? (Launch Failure)

USA Pioneer 1 (11 Oct 1958)
Attempted Lunar Orbit (Launch Failure)

USSR Luna 1958B (12 Oct 1958)
Attempted Lunar Impact? (Launch Failure)

USA Pioneer 2 (8 Nov 1958)
Attempted Lunar Orbit (Launch Failure)

USSR Luna 1958C (4 Dec 1958)
Attempted Lunar Impact? (Launch Failure)

USA Pioneer 3 (6 Dec 1958)
Attempted Lunar Flyby (Launch Failure)

1959
USSR Luna 1 (2 Jan 1959)
Lunar Flyby (Attempted Lunar Impact?)

USA Pioneer 4 (3 Mar 1959)
Lunar Flyby

USSR Luna 1959A (16 Jun 1959)
Attempted Lunar Impact? (Launch Failure)

USSR Luna 2 (12 Sep 1959)
Lunar Impact

USSR Luna 3 (4 Oct 1959) Lunar Flyby

USA Pioneer P-3 (26 Nov 1959)
Attempted Lunar Orbiter (Launch Failure)

1960
USSR Luna 1960A (15 Apr 1960)
Attempted Lunar Flyby (Launch Failure)

USSR Luna 1960B (18 Apr 1960)
Attempted Lunar Flyby (Launch Failure)

USA Pioneer P-30 (25 Sep 1960)
Attempted Lunar Orbiter (Launch Failure)

USSR Marsnik 1 (Mars 1960A)
(10 Oct 1960) Attempted Mars Flyby
(Launch Failure)

USSR Marsnik 2 (Mars 1960B)
(14 Oct 1960) Attempted Mars Flyby
(Launch Failure)

USA Pioneer P-31 (15 Dec 1960)
Attempted Lunar Orbiter (Launch Failure)

1961
USSR Sputnik 7 (4 Feb 1961)
Attempted Venus Impact

USSR Venera 1 (12 Feb 1961)
Venus Flyby (Contact Lost)

USA Ranger 1 (23 Aug 1961)
Attempted Lunar Test Flight

USA Ranger 2 (18 Nov 1961)
Attempted Lunar Test Flight

1962
USA Ranger 3 (26 Jan 1962)
Attempted Lunar Impact

USA Ranger 4 (23 Apr 1962)
Lunar Impact (Contact Lost)

USA Mariner 1 (22 Jul 1962)
Attempted Venus Flyby (Launch Failure)

USSR Sputnik 23 (25 Aug 1962)
Attempted Venus Flyby

USA Mariner 2 (27 Aug 1962)
Venus Flyby

USSR Sputnik 24 (1 Sep 1962)
Attempted Venus Flyby

USSR Sputnik 25 (12 Sep 1962)
Attempted Venus Flyby

USA Ranger 5 (18 Oct 1962)
Attempted Lunar Impact

USSR Sputnik 29 (24 Oct 1962)
Attempted Mars Flyby

USSR Mars 1 (1 Nov 1962)
Mars Flyby (Contact Lost)

USSR Sputnik 31 (4 Nov 1962)
Attempted Mars Flyby

1963
USSR Sputnik 33 (4 Jan 1963)
Attempted Lunar Lander

USSR Luna 1963B (2 Feb 1963)
Attempted Lunar Lander (Launch Failure)

USSR Luna 4 (2 Apr 1963)
Attempted Lunar Lander

USSR Cosmos 21 (11 Nov 1963)
Attempted Venera Test Flight?

1964
USA Ranger 6 (30 Jan 1964)
Lunar Impact (Cameras Failed)

USSR Venera 1964A (19 Feb 1964)
Attempted Venus Flyby (Launch Failure)

USSR Venera 1964B (1 Mar 1964)
Attempted Venus Flyby (Launch Failure)

USSR Luna 1964A (21 Mar 1964)
Attempted Lunar Lander (Launch Failure)

USSR Cosmos 27 (27 Mar 1964)
Attempted Venus Flyby

USSR Zond 1 (2 Apr 1964)
Venus Flyby (Contact Lost)

USSR Luna 1964B (20 Apr 1964)
Attempted Lunar Lander (Launch Failure)

USSR Zond 1964A (4 Jun 1964)
Attempted Lunar Lander (Launch Failure)

USA Ranger 7 (28 Jul 1964)
Lunar Impact

USA Mariner 3 (5 Nov 1964)
Attempted Mars Flyby

USA Mariner 4 (28 Nov 1964)
Mars Flyby

USSR Zond 2 (30 Nov 1964)
Mars Flyby (Contact Lost)

1965
USA Ranger 8 (17 Feb 1965)
Lunar Impact

USSR Cosmos 60 (12 Mar 1965)
Attempted Lunar Lander

USA Ranger 9 (21 Mar 1965)
Lunar Impact

USSR Luna 1965A (10 Apr 1965)
Attempted Lunar Lander? (Launch Failure)

USSR Luna 5 (9 May 1965)
Lunar Impact (Attempted Soft Landing)

USSR Luna 6 (8 Jun 1965)
Attempted Lunar Lander

USSR Zond 3 (18 Jul 1965) Lunar Flyby

USSR Luna 7 (4 Oct 1965)
Lunar Impact (Attempted Soft Landing)

USSR Venera 2 (12 Nov 1965)
Venus Flyby (Contact Lost)

USSR Venera 3 (16 Nov 1965)
Venus Lander (Contact Lost)

USSR Cosmos 96 (23 Nov 1965)
Attempted Venus Lander?

USSR Venera 1965A (23 Nov 1965)
Attempted Venus Flyby (Launch Failure)

USSR Luna 8 (3 Dec 1965)
Lunar Impact (Attempted Soft Landing?)

1966
USSR Luna 9 (31 Jan 1966) Lunar Lander

USSR Cosmos 111 (1 Mar 1966)
Attempted Lunar Orbiter?

USSR Luna 10 (31 Mar 1966)
Lunar Orbiter

USSR Luna 1966A (30 Apr 1966)
Attempted Lunar Orbiter? (Launch Failure)

USA Surveyor 1 (30 May 1966)
Lunar Lander

USA Explorer 33 (1 Jul 1966)
Attempted Lunar Orbiter

USA Lunar Orbiter 1 (10 Aug 1966)
Lunar Orbiter

USSR Luna 11 (24 Aug 1966)
Lunar Orbiter

USA Surveyor 2 (20 Sep 1966)
Attempted Lunar Lander

USSR Luna 12 (22 Oct 1966)
Lunar Orbiter

USA Lunar Orbiter 2 (6 Nov 1966)
Lunar Orbiter

USSR Luna 13 (21 Dec 1966)
Lunar Lander

1967
USA Lunar Orbiter 3 (4 Feb 1967)
Lunar Orbiter

USA Surveyor 3 (17 Apr 1967)
Lunar Lander

USA Lunar Orbiter 4 (8 May 1967)
Lunar Orbiter

USSR Venera 4 (12 Jun 1967)
Venus Probe

USA Mariner 5 (14 Jun 1967) Venus Flyby

USSR Cosmos 167 (17 Jun 1967)
Attempted Venus Probe

USA Surveyor 4 (14 Jul 1967)
Attempted Lunar Lander

USA Explorer 35 (IMP-E) (19 Jul 1967)
Lunar Orbiter

USA Lunar Orbiter 5 (1 Aug 1967)
Lunar Orbiter

USA Surveyor 5 (8 Sep 1967)
Lunar Lander

USSR Zond 1967A (28 Sep 1967)
Attempted Lunar Test Flight (Launch Failure)

USA Surveyor 6 (7 Nov 1967)
Lunar Lander

USSR Zond 1967B (22 Nov 1967)
Attempted Lunar Test Flight (Launch Failure)

1968
USA Surveyor 7 (7 Jan 1968) Lunar Lander

USSR Luna 1968A (7 Feb 1968)
Attempted Lunar Orbiter (Launch Failure)

USSR Zond 4 (2 Mar 1968) Test Flight

USSR Luna 14 (7 Apr 1968) Lunar Orbiter

USSR Zond 1968A (23 Apr 1968)
Attempted Lunar Test Flight? (Launch Failure)

USSR Zond 5 (15 Sep 1968)
Lunar Flyby and Return to Earth

USSR Zond 6 (10 Nov 1968)
Lunar Flyby and Return to Earth

USA Apollo 8 (21 Dec 1968)
Manned Lunar Orbiter

1969
USSR Venera 5 (5 Jan 1969) Venus Probe

USSR Venera 6 (10 Jan 1969)
Venus Probe

USSR Zond 1969A (20 Jan 1969)
Attempted Lunar Flyby and Return (Launch Failure)

USSR Luna 1969A (19 Feb 1969)
Attempted Lunar Rover? (Launch Failure)

USSR Zond L1S-1 (21 Feb 1969)
Attempted Lunar Orbiter (Launch Failure)

USA Mariner 6 (25 Feb 1969) Mars Flyby

USA Mariner 7 (27 Mar 1969) Mars Flyby

USSR Mars 1969A (27 Mar 1969)
Attempted Mars Lander? (Launch Failure)

USSR Mars 1969B (2 Apr 1969)
Attempted Mars Lander? (Launch Failure)

USSR Luna 1969B (15 Apr 1969)
Attempted Lunar Sample Return?
(Launch Failure)

USA Apollo 10 (18 May 1969)
Manned Lunar Orbiter

USSR Luna 1969C (14 Jun 1969)
Attempted Lunar Sample Return?
(Launch Failure)

USSR Zond L1S-2 (3 Jul 1969)
Attempted Lunar Orbiter
(Launch Failure)

USSR Luna 15 (13 Jul 1969)
Attempted Lunar Sample Return

USA Apollo 11 (16 Jul 1969)
Manned Lunar Landing

USSR Zond 7 (7 Aug 1969)
Lunar Flyby and Return to Earth

USSR Cosmos 300 (23 Sep 1969)
Attempted Lunar Sample Return?

USSR Cosmos 305 (22 Oct 1969)
Attempted Lunar Sample Return?

USA Apollo 12 (14 Nov 1969)
Manned Lunar Landing

1970
USSR Luna 1970A (6 Feb 1970)
Attempted Lunar Sample Return?
(Launch Failure)

USSR Luna 1970B (19 Feb 1970)
Attempted Lunar Orbiter? (Launch
Failure)

USA Apollo 13 (11 Apr 1970)
Manned Lunar Mission (Landing
Aborted)

USSR Venera 7 (17 Aug 1970)
Venus Lander

USSR Cosmos 359 (22 Aug 1970)
Attempted Venus Probe

USSR Luna 16 (12 Sep 1970)
Lunar Sample Return

USSR Zond 8 (20 Oct 1970)
Lunar Flyby and Return to Earth

USSR Luna 17/Lunokhod 1
(10 Nov 1970) Lunar Rover

1971
USA Apollo 14 (31 Jan 1971)
Manned Lunar Landing

USA Mariner 8 (8 May 1971)
Attempted Mars Orbiter (Launch Failure)

USSR Cosmos 419 (10 May 1971)
Attempted Mars Orbiter/Lander

USSR Mars 2 (19 May 1971)
Mars Orbiter/ Attempted Lander

USSR Mars 3 (28 May 1971)
Mars Orbiter/ Lander

USA Mariner 9 (30 May 1971)
Mars Orbiter

USA Apollo 15 (26 Jul 1971)
Manned Lunar Landing

USSR Luna 18 (2 Sep 1971)
Attempted Lunar Sample Return

USSR Luna 19 (28 Sep 1971)
Lunar Orbiter

1972
USSR Luna 20 (14 Feb 1972)
Lunar Sample Return

USA Pioneer 10 (3 Mar 1972)
Jupiter Flyby

USSR Venera 8 (27 Mar 1972)
Venus Lander

USSR Cosmos 482 (31 Mar 1972)
Attempted Venus Probe

USA Apollo 16 (16 Apr 1972)
Lunar Manned Landing

USSR Soyuz L3 (23 Nov 1972)
Attempted Lunar Orbiter (Launch Failure)

USA Apollo 17 (7 Dec 1972)
Lunar Manned Landing

1973
USSR Luna 21/Lunokhod 2 (8 Jan 1973)
Lunar Rover

USA Pioneer 11 (5 Apr 1973)
Jupiter/Saturn Flyby

USA Explorer 49 (RAE-B) (10 Jun 1973)
Lunar Orbiter/Radio Astronomy

USSR Mars 4 (21 Jul 1973)
Mars Flyby (Attempted Mars Orbiter)

USSR Mars 5 (25 Jul 1973) Mars Orbiter

USSR Mars 6 (5 Aug 1973)
Mars Lander (Contact Lost)

USSR Mars 7 (9 Aug 1973)
Mars Flyby (Attempted Mars Lander)

USA Mariner 10 (4 Nov 1973)
Venus/Mercury Flybys

1974
USSR Luna 22 (2 Jun 1974)
Lunar Orbiter

USSR Luna 23 (28 Oct 1974)
Attempted Lunar Sample Return

1975
USSR Venera 9 (8 Jun 1975)
Venus Orbiter and Lander

USSR Venera 10 (14 Jun 1975)
Venus Orbiter and Lander

USA Viking 1 (20 Aug 1975)
Mars Orbiter and Lander

USA Viking 2 (9 Sep 1975)
Mars Orbiter and Lander

USSR Luna 1975A (16 Oct 1975)
Attempted Lunar Sample Return?

1976
USSR Luna 24 (9 Aug 1976)
Lunar Sample Return

1977
USA Voyager 2 (20 Aug 1977)
Jupiter/Saturn/Uranus/Neptune Flyby

USA Voyager 1 (5 Sep 1977)
Jupiter/Saturn Flyby

1978
USA Pioneer Venus 1 (20 May 1978)
Venus Orbiter

USA Pioneer Venus 2 (8 Aug 1978)
Venus Probes

USA ISEE-3/ICE (12 Aug 1978)
Comet Giacobini-Zinner and Halley
Flybys

USSR Venera 11 (9 Sep 1978)
Venus Flyby and Lander

USSR Venera 12 (14 Sep 1978)
Venus Flyby and Lander

1981
USSR Venera 13 (30 Oct 1981)
Venus Orbiter and Lander

USSR Venera 14 (4 Nov 1981)
Venus Orbiter and Lander

1983
USSR Venera 15 (2 Jun 1983)
Venus Orbiter

USSR Venera 16 (7 Jun 1983)
Venus Orbiter

1984
USSR Vega 1 (15 Dec 1984)
Venus Lander and Balloon/Comet Halley
Flyby

USSR Vega 2 (21 Dec 1984)
Venus Lander and Balloon/Comet Halley
Flyby

1985
Japan Sakigake (7 Jan 1985)
Comet Halley Flyby

ESA Giotto (2 Jul 1985)
Comet Halley Flyby

Japan Suisei (Planet-A) (18 Aug 1985)
Comet Halley Flyby

1988
USSR Phobos 1 (7 Jul 1988)
Attempted Mars Orbiter/Phobos Landers

USSR Phobos 2 (12 Jul 1988)
Mars Orbiter/Attempted Phobos Landers

1989
USA Magellan (4 May 1989)
Venus Orbiter

USA Galileo (18 Oct 1989)
Jupiter Orbiter/Probe

1990
Japan Hiten (24 Jan 1990)
Lunar Flyby and Orbiter

ESA Ulysses (06 Oct 1990)
Jupiter Flyby and Solar Polar Orbiter

1992
USA Mars Observer (25 Sep 1992)
Attempted Mars Orbiter (Contact Lost)

1994
USA Clementine (25 Jan 1994)
Lunar Orbiter/Attempted Asteroid Flyby

1996
USA NEAR (17 Feb 1996)
Asteroid Eros Orbiter

USA Mars Global Surveyor
(7 Nov 1996) Mars Orbiter

CIS (former USSR) Mars 96
(16 Nov 1996) Attempted Mars

Orbiter/Landers (Launch Failure)

USA Mars Pathfinder (4 Dec 1996)
Mars Lander and Rover

1997
USA Cassini (15 Oct 1997) Saturn Orbiter

ESA Huygens (15 Oct 1997) Titan Probe

USA AsiaSat 3/HGS-1 (24 Dec 1997)
Lunar Flyby

1998
USA Lunar Prospector (7 Jan 1998)
Lunar Orbiter

Japan Nozomi (Planet-B) (3 Jul 1998)
Mars Orbiter

USA Deep Space 1 (DS1) (24 Oct 1998)
Asteroid and/or Comet Flyby

USA Mars Climate Orbiter (11 Dec 1998)
Mars Orbiter

1999
USA Mars Polar Lander (3 Jan 1999)
Mars Lander

USA Deep Space 2 (DS2) (3 Jan 1999)
Mars Penetrator

USA Stardust (6 Feb 1999)
Comet Coma Sample Return

Japan Lunar-A (Summer 1999)
Lunar Orbiter and Penetrators

2001
USA Genesis (Jan 2001)
Solar Wind Sample Return

USA Mars Surveyor 2001 Orbiter
(7 Mar 2001) Mars Orbiter

USA Mars Surveyor 2001 Lander
(5 Apr 2001) Mars Lander/Rover

2002
Japan Muses-C (Jan 2002)
Asteroid Lander and Sample Return

USA CONTOUR (Jul 2002)
Flyby of three Comet Nuclei

2003
ESA Rosetta (23 Jan 2003)
Comet Orbiter and Lander

USA Champollion/DS4 (May 2003)
Comet Sample Return

USA Mars Surveyor 2003 (May 2003)
Mars Orbiter and Lander

ESA Mars Express (Jun 2003)
Mars Orbiter and Lander

USA Europa Orbiter (2003)
Proposed Europa Orbiter

Japan Selene (2003) Lunar Orbiter

2004
USA Pluto-Kuiper Express (Dec 2004)
Proposed Flyby of Pluto and Kuiper Belt

2006
USA Mars Surveyor 2005 (Jul 2005)
Mars Orbiter and Lander

The planets: the statistics

PLANET	Diameter at equator (km and relative to Earth)	Mass relative to Earth	Density relative to water	Volume relative to Earth	Average temperature degrees Centigrade (S = surface, C = clouds)	Surface gravity relative to Earth	Atmosphere (main components)
Mercury	4,878 0.38	0.055	5.5	0.06	350(S) day -170(S) night	0.38	None
Venus	12,103 0.95	0.81	5.2	0.88	-33(C) day 480(S) night	0.9	Carbon dioxide
Earth	12,756 1.00	1.00	5.5	1.00	22(S)	1.0	Nitrogen, oxygen
Mars	6786 0.53	0.11	3.9	0.15	23(S)	0.38	Carbon dioxide, argon
Jupiter	142,980 11.0	318	1.3	1,316	-150(C)	2.64	Hydrogen, helium
Saturn	120,540 9.41	95	0.7	755	-180(C)	1.16	Hydrogen, helium
Uranus	51,120 4.11	15	1.3	67	-210(C)	1.11	Hydrogen, helium, methane
Neptune	49,530 3.96	17	1.6	57	-220(C)	1.21	Hydrogen, helium, methane
Pluto	2,280 0.18	0.002	2.1	0.015(?)	-230(?)	0.06	Nitrogen?

PLANET	Tilt of axis (degrees)	Inclination of orbit to ecliptic (degrees)	Eccentricity of orbit	Period of revolution (year)	Period of rotation[1]	Mean distance from Sun (millions of km)	Number of known moons
Mercury	0.0(?)	7	0.206	88d	59d (176d)	57.9	0
Venus	2	3.4	0.007	225d	243d (E to W) (117d)	108.2	0
Earth	23.27	0	0.017	365d	23hr 56m (24hr)	149.6	1
Mars	24.46	1.9	0.093	687d	24hr 37m	227.9	2
Jupiter	3.05	1.3	0.048	11.9yr	9hr 55m	778.3	16
Saturn	26.44	2.5	0.056(?)	29.5yr	10hr 39m	1,427	18
Uranus	97.53	0.8	0.047	84yr	17hr 14m (E to W)	2,869.6	15
Neptune	28.48	1.8	0.009	165yr	16hr 7m	4,496.6	8
Pluto	?	17.2	0.25	248yr	6d 9hr (E to W)	5,900	1

[1] Relative to the stars. If the length of the 'day' (sunrise to sunrise) is very different, it is given in brackets.

PLANET	Mean orbital velocity (km per second)	Oblateness
Mercury	47.9	0
Venus	35	0
Earth	29.8	0.003
Mars	24.1	0.009
Jupiter	13.1	0.06
Saturn	9.7	0.1
Uranus	6.8	0.06
Neptune	5.4	0.02
Pluto	4.7	?

Moons of the planets

MOON	Mean distance from centre of planet (km)	Orbital period (days)	Diameter (km)	MOON	Mean distance from centre of planet (km)	Orbital period (days)	Diameter (km)
				Dione	377,420	2.737	1,120
EARTH				Rhea	527,040	4.518	1,528
				Titan	1,221,860	15.945	5,150
Moon	384,400	27.321	3475.6	Hyperion	1,481,100	21.277	360 x 280 x 225
				Iapetus	3,651,300	79.331	1,436
MARS				Phoebe	12,954,000	550.4*	230 x 220 x 210
Phobos	9,270	0.32	17 x 14 x 11				
Deimos	23,400	1.26	9 x 7 x 6	**URANUS**			
				Cordelia	49,471	0.330	26
JUPITER				Ophelia	53,796	0.372	30
Metis	127,960	0.295	40	Bianca	59,173	0.433	42
Adrastea	128,980	0.298	26 x 20 x 16	Cressida	61,777	0.463	62
Amalthea	181,300	0.498	262 x 146 x 143	Desdemona	62,676	0.475	54
Thebe	221,900	0.675	68 x 56	Juliet	64,372	0.493	84
Io	421,600	1.769	3,660 x 3,637 x 3,631	Portia	66,085	0.513	108
Europa	670,900	3.551	3,130	Rosalind	69,941	0.558	54
Ganymede	1,070,000	7.155	5,268	Belinda	75,258	0.622	66
Callisto	1,880,000	16.689	4,806	Puck	86,000	0.762	154
Leda	11,094,000	238.7	10	Miranda	129,400	1.414	472
Himalia	11,480,000	250.6	170	Ariel	191,000	2.520	1,158
Lysithea	11,720,000	259.2	24	Umbriel	266,300	4.144	1,169
Elara	11,737,000	259.7	80	Titania	435,000	8.706	1,578
Ananke	21,200,000	631*	20	Oberon	583,500	13.463	1,523
Carme	22,600,000	692*	30				
Pasiphaë	23,500,000	735*	36	**NEPTUNE**			
Sinope	23,700,000	758*	28	Naiad	48,000	0.296	54
				Thalassa	50,000	0.312	80
SATURN				Despina	52,500	0.333	180
Pan	133,600	0.57	12	Galatea	62,000	0.429	150
Atlas	137,670	0.602	37 x 34 x 27	Larissa	73,600	0.554	192
Prometheus	139,350	0.613	148 x 100 x 68	Proteus	117,600	1.121	416
Pandora	141,700	0.629	110 x 88 x 62	Triton	354,800	5.877*	2,705
Epimetheus	151,420	0.694	194 x 190 x 154	Nereid	1,345,500 – 9,688,500	360.16	240
Janus	151,470	0.695	138 x 110 x 110				
Mimas	185,540	0.942	421 x 395 x 385	**PLUTO**			
Enceladus	238,040	1.370	512 x 495 x 488	Charon	19,640	6.387	1,212
Tethys	294,670	1.888	1,046				
Telesto	294,670	1.888	30 x 25 x 15	(*retrograde)			
Calypso	294,670	1.888	30 x 16 x 16				
Helene	377,410	2.737	35				

Extrasolar planets

Planets found orbiting other stars to date

Star	Mass of planet (Jupiter=1)	Distance of planet from star (Earth to Sun=1)	Rotation period of planet around star
51 Pegasi	0.47	0.05	4.2 days
HD187123	0.52	0.04	3.1 days
Upsilon Andromeda	0.68	0.06	4.6 days
55 Cancri	1.9	0.11	14.6 days
Rho Corona Borealis	1.1	0.23	39.6 days
HD217107	1.28	0.04	7.1 days
HD210277	1.37	1.15	1.2 years
16 Cygni B	1.5	0.6-2.7 (oval-shaped orbit)	2.2 years
Gliese 876	2.1	0.21	60.8 days
47 Ursae Majoris	2.8	2.11	2.98 years
14 Herculis	3.3	2.5	4.43 years
HD195019	3.43	0.14	18.3 days
Tau Bootis	3.87	0.04	3.3 days
HD168443	4.1	0.2	70 days
Gliese 86	4.9	0.11	15.8 days
70 Virginis	6.6	0.43	117 days
HD114762	10	0.3	84 days

Further reading

Chapter 1
David H. Levy, *Clyde Tombaugh: Discoverer of Planet Pluto*, University of Arizona Press, 1991
Stuart Ross Taylor, *Solar System Evolution: A New Perspective*, Cambridge University Press, 1992
Alan Stern and Jacqueline Mitton, *Pluto and Charon*, John Wiley & Sons, 1998

Chapter 2
Buzz Aldrin and Malcolm McConnell, *Men from Earth*, Bantam, 1989
Andrew Chaikin, *A Man on the Moon*, Penguin, 1994
Brian Harvey, *The New Russian Space Programme*, John Wiley & Sons, 1996
Walter A. McDougall, *The Heavens and the Earth: A Political History of the Space Age*, Johns Hopkins University Press, 1997
James E. Oberg, *Red Star in Orbit*, Random House, 1981
Alan Shepard and Deke Slayton, *Moon Shot: The Inside Story of America's Race to the Moon*, Turner Publishing Inc., 1994
Don E. Wilhelms, *To a Rocky Moon: A Geologist's History of Lunar Exploration*, University of Arizona Press, 1993
Tom Wolfe, *The Right Stuff*, Picador, 1991

Chapter 3
Jay Barbree, Martin Caidim and Susan Wright, *Destination Mars*, Penguin Studio, 1997
Henry S. F. Cooper, *The Evening Star: Venus Observed*, Johns Hopkins University Press, 1994
Charles Frankel, *Volcanoes of the Solar System*, Cambridge University Press, 1996
David Harry Grinspoon, *Venus Revealed*, Addison-Wesley, 1997

Chapter 4
Mark Littman, *Planets Beyond: Discovering the Outer Solar System*, John Wiley & Sons, 1998
David W. Swift, *Voyager Tales: Personal Views of the Grand Tour*, American Institute of Aeronautics and Astronautics Inc., 1997

Chapter 5
John Gribbin, *Companion to the Cosmos*, Phoenix, 1997

Rudolph Kippenhaan, *Discovering the Secrets of the Sun*, John Wiley & Sons, 1994
Kenneth R. Lang, *Sun, Earth and Sky*, Springer, 1997
Kenneth J. H. Phillips, *Guide to the Sun*, Cambridge University Press, 1992

Chapter 6
Craig Ryan, *The Pre-Astronauts: Manned Ballooning on the Threshold of Space*, Naval Institute Press, 1995

Chapter 7
Harry Y. McSween, Jr., *Stardust to Planets: A Geological Tour of the Universe*, St Martin's Griffin, 1995
Carl Sagan, *Cosmos*, Abacus, 1985
Robert Zubrin, *The Case for Mars*, Free Press, 1996

Chapter 8
Ken Croswell, *Planet Quest: The Epic Discovery of Alien Solar Systems*, Harcourt Brace, 1998
Paul Halpern, *The Quest for Alien Planets: Exploring Worlds Outside the Solar System*, Plenum Press, 1997
David Malin, *A View of the Universe*, Cambridge University Press, 1993
Carl Sagan, *Pale Blue Dot: A Vision of the Human Future in Space*, Headline, 1995

General
Jay Apt, Michael Helfert and Justin Wilkinson, *Orbit: NASA Astronauts Photograph the Earth*, National Geographic Society, 1996
J. Kelley Beatty, Carolyn Collins Petersen and Andrew Chaikin, *The New Solar System*, Cambridge University Press, 1998
Andrew Franknol, David Morrison and Sydney C. Wolff, *Voyages through the Universe*, Saunders College Publishing, 1997
Ronald Greeley and Raymond Batson, *NASA Atlas of the Solar System*, Cambridge University Press, 1997
Bruce Murray, *Journey Into Space: The First Thirty Years of Space Exploration*, Norton, 1990

Websites

General Space

www.nasa.gov/ – The space site: welcome to NASA.

www.iki.rssi.ru – The Russian Space Agency.

www.jsc.nasa.gov – Johnson Space Center. Information about NASA manned missions.

planetary.org – The Planetary Society pages.

pds.jpl.nasa.gov/ – NASA's planetary data.

bang.lanl.gov/solarsys/ – Solar system pictures and historical astronomy.

photojournal.jpl.nasa.gov – All the photos of planets and moons NASA's probes have ever taken.

seds.lpl.arizona.edu/ – Students for the exploration and development of space.

www.esrin.esa.it/ – European Space Agency – includes links to lots of other space agencies.

www.stsci.edu/ – Hubble Space telescope site.

pathfinder.com/Life/space/giantleap/index.html – *Life* magazine reports and photographs of major events in space exploration. Particularly good features on the *Mercury, Gemini* and *Apollo* missions.

antwrp.gsfc.nasa.gov/apod/astropix.html – Astronomy picture of the day. Each day a different image or photograph is featured, with a brief explanation written by a professional astronomer.

www.sci.esa.int – European Space Agency's science page.

Encyclopaedias

nssdc.gsfc.nasa.gov/ – Chronology of all space missions.

solar.rtd.utk.edu:81/~mwade/spaceflt.htm – Comprehensive chronology of space flight.

leonardo.jpl.nasa.gov/msl/home.html – Mission and spacecraft library.

News

www.hq.nasa.gov/office/pao/NewsRoom/today.html – NASA daily news.

www.hq.nasa.gov/osf/ – General Space News (NASA).

Education sites

spacelink.nasa.gov – Space resources for educators site.

ceps.nasm.edu/ – National Air and Space Museum.

www.ccas.ru/~chernov/vsm/main.htm – Russian Spacecraft Museum: lots of rare pictures.

windows.ivv.nasa.gov/ – NASA's windows on the Universe site.

www.soest.hawaii.edu/PSRdiscoveries/ – Cutting edge and accessible planetary science.

www.sji.org – San Juan Capistrano astronomy newsletter.

Extrasolar Planets

seds.lpl.arizona.edu/billa/tnp/other.html – A summary of discoveries so far and links to other related sites.

www.obspm.fr/departement/darc/planets/encycl.html – Encyclopedia of extrasolar planet discoveries.

Sun

sohowww.nascom.nasa.gov/ – *SOHO* mission web site.

www.bbso.njit.edu/ – Big Bear solar observatory.

www.astro.ucla.edu/~obs/intro.html – Mt. Wilson's solar observatory, daily pictures of the Sun.

ulysses.jpl.nasa.gov/ULSHOME/ulshome.html – *Ulysees* Solar probe.

Moon

www.jsc.nasa.gov/pao/apollo17/ – Interview with *Apollo 17* astronauts.

nssdc.gsfc.nasa.gov/planetary/lunar/ – Information on all lunar missions.

www.hq.nasa.gov/alsj/ – *Apollo* lunar surface journal. Transcripts of all communication between Houston mission control and *Apollo* Moonwalkers.

lunarprospector.arc.nasa.gov – *Lunar Prospector*.

Venus

cass.jsc.nasa.gov/publications/slidesets/venus.html – Background and slide set.

nssdc.gsfc.nasa.gov/photo_gallery/photogallery-venus.html – 3-D pictures.

www.jpl.nasa.gov/magellan/ – *Magellan* mission to Venus.

Mars

www-pdsimage.wr.usgs.gov/PDS/public/mapmaker/mapmkr.htm – Armchair tour of Mars.

www.fas.org/mars – Federation of American Scientists. Lots of Mars links.

www.jpl.nasa.gov/mars – All NASA's upcoming Mars missions and *Mars Pathfinder*.

Jupiter

www.isc.tamu.edu/~astro/jupiter.html – Jupiter events and links to lots of other Jupiter sites.

www.jpl.nasa.gov/galileo/ – *Galileo* mission to Jupiter and its moons.

ccf.arc.nasa.gov/galileo_probe – *Galileo* probe into Jupiter.

Saturn

ringmaster.arc.nasa.gov – The ultimate pictures of Saturn's rings.

www.jpl.nasa.gov/cassini/ – *Cassini* mission to Saturn and its moons.

www.estec.esa.nl/spdwww/huygens/ – *Huygens* mission to Titan.

Other probe missions

nmp.jpl.nasa.gov/ – NASA's New Millennium Programme.

quest.arc.nasa.gov/pioneer10/ – *Pioneer 10* and *11* 25th-anniversary site.

www.jpl.nasa.gov/mip/voyager.html – *Voyager 1* and *2*.

stardust.jpl.nasa.gov/ – *Stardust* mission to return a comet sample.

www.sci.esa.int/rosetta – *Rosetta* mission to a comet.

sd-www.jhuapl.edu/NEAR/ – *NEAR* mission to an asteroid.

Picture credits

BBC Books would like to thank the following for providing photographs and for permission to reproduce copyright material. While every effort has been made to trace and acknowledge all copyright holders, we would like to apologise for any errors or omissions.

Key: a above, b below, r right, l left, c centre

Abbreviations: NASA — National Aeronautics and Space Administration
SPL — Science Photo Library
JPL — NASA Jet Propulsion Laboratory

2 BBC; 6 Russian State Archive of Science and Technology; 8 BBC; 11l Dr Chris Riley; 11r Lowell Observatory; 12l Lowell Observatory; 12r Dr Chris Riley; 13l Dr Chris Riley; 13r Dr Chris Riley; 14 Galaxy Picture Library; 15 Space Telescope Science Institute/NASA/SPL; 16l NASA; 16r Photovault; 17l United States Geological Survey/SPL; 17r Galaxy Picture Library; 18al Space Telescope Science Institute/NASA/SPL; 18ar NASA/Genesis Space Photo Library; 18bl NASA/SPL; 19 NASA/SPL; 21 Space Telescope Institute/NASA/SPL; 22al NASA/Genesis Space Photo Library; 22bl SPL; 22br AKG London; 23 NASA/SPL; 24al AKG London; 24bl NASA/Genesis Space Photo Library; 24br Imperial War Museum; 25 Natural History Museum; 26 NASA/Galaxy Picture Library; 27a NASA/Galaxy Picture Library; 27b NASA; 28 Francois Gohier/SPL; 30 David Hardy/SPL; 31 NASA/SPL; 33 background JPL; 33ar JPL; 33br JPL; 35ar European Space Agency/SPL; 35c AKG London; 35br Jerry Lodrigoss/SPL; 36 White Sands Missile Range; 37 Simon Fraser/SPL; 38 NASA/SPL; 40 Scala; 41 Image Select; 42 Spacecharts; 44 John Sanford/SPL; 45 NASA; 47l JPL; 47r NASA; 48 JPL; 49a NASA/SPL; 49b Russian State Archive of Science and Technology; 50 John Frost; 51l Russian State Archive of Science and Technology; 51r John Frost; 52l NASA/Genesis Space Photo Library; 52r NASA; 53 NASA; 54 Galaxy Picture Library; 55 NASA; 56al Novosti/SPL; 56bl NASA; 56br NASA; 58 NASA/Genesis Space Photo Library; 59 NASA; 60l Spacecharts; 60c Spacecharts; 60r NASA/SPL; 61l NASA/SPL; 61c NASA; 61r Spacecharts; 62 NASA; 63l Russian State Archive of Science and Technology; 63r Russian State Archive of Science and Technology; 64 Genesis Space Photo Library; 65 NASA; 66a NASA/SPL; 66b NASA; 67 NASA; 68 NASA/SPL; 69 NASA; 70l NASA/SPL; 70r Genesis Space Photo Library; 71r NASA/SPL; 72 BBC; 73a Ballistic Missile Defense Organization/Naval Research Laboratory/Lawrence Livermore National Laboratory/SPL; 73b JPL; 75 NASA/SPL; 77l Russian State Archive of Science and Technology; 77r NASA; 78l National Geographic Society; 78r United States Geological Survey/SPL; 79 National Geographic Society; 80 United States Geological Survey/SPL; 81l Peter Menzell/SPL; 81c Professor Stewart Lowther/SPL; 81r Planet Earth Pictures/Robert Hessler; 82 NASA/SPL; 83 NASA/SPL; 84 NASA/SPL; 85a NASA/SPL; 85b NASA/SPL; 86 NASA/SPL; 87 JPL; 89 NASA; 89r Planet Earth Pictures; 90 Planet Earth Pictures; 91a Royal Astronomical Society; 92 JPL; 93al JPL; 93ar JPL; 93b JPL; 94 JPL; 94b Planet Earth Pictures; 95 United States Geological Survey/SPL; 96 JPL; 97 United States Geological Survey/SPL; 98 NASA/SPL; 99 JPL; 100 JPL; 101a JPL; 101b Malin Space Sciences NASA; 102a JPL; 102b Genesis Space Photo Library; 103 NASA/SPL; 104 JPL; 106 Space Telescope Science Institute/NASA/SPL; 108 NASA; 109 Genesis Space Photo Library; 110 NASA/Galaxy Picture Library; 111b Galaxy Picture Library; 112 National Radio Astronomy Observatory/Associated Universities Inc./SPL; 114 Colorific!; 115 Royal Astronomical Society; 116a NASA; 116b JPL; 117a Royal Greenwich Observatory/SPL; 117b ET Archive; 119al NASA; 119ar Genesis Space Photo Library; 119b BBC; 120 NASA/SPL; 121l Planet Earth Pictures; 121c NASA/SPL; 121r NASA/SPL; 122a Galaxy Picture Library; 122b JPL; 124a Genesis Space Photo Library; 124b NASA/SPL; 125a Galaxy Picture Library; 125c NASA/SPL; 125b Galaxy Picture Library; 126a NASA/SPL; 126b SPL; 127a NASA/SPL; 127b NASA/SPL; 128a NASA/SPL; 128b Space Telescope Science Institute/NASA/SPL; 129a United States Geological Survey; 129c NASA/SPL; 129b NASA/SPL; 130 Genesis Space Photo Library; 131 Julian Baum/SPL; 132 NASA/SPL; 133ar Image Select; 133bl L'Illustration/Sygma; 133br SPL; 134a David Parker/SPL; 134b NASA/SPL; 135a NASA/SPL; 135b JPL; 136 NASA; 138 European Space Agency/SPL; 139 ET Archive; 140 ET Archive; 141 Hulton Getty Picture Collection; 142 National Solar Observatory Sacremento Peak; 143l ET Archive; 143r Artephot/A.D.P.C.; 144a David Leah/SPL; 144b ET Archive; 145ar R.Muller Observatoire Midi-Pyrénées; 145l R.Muller/Observatoire Midi-Pyrénées; 145 SPL; 146 California Institute of Technology; 147 NASA; 148 Planet Earth Pictures; 149 AKG London; 150 United States Department of Energy/SPL; 151a Royal Astronomical Society; 151b Royal Astronomical Society; 152 BBC; 153 Planet Earth Pictures; 154 Newton Magazine, the Kyoikusha Company; 155 Yi-Ming Wang; 156 Planet Earth Pictures; 157a Genesis Space Photo Library; 157b Richard J. Wainscoat, Peter Arnold Inc/SPL; 158 European Space Agency/SPL; 160 NASA; 161l NASA/SPL; 161r NASA/SPL; 162a NASA; 162b NASA; 163 NASA; 164 NASA/SPL; 165 European Space Agency; 166 Galaxy Picture Library; 169l Genesis Space Photo Library; 169r Major David G Simons; Life Magazine ©Time Inc/Katz Pictures; 170 NASA; 171 NASA; 172 Photovault; 173 James Younger; 174l Galaxy Picture Library; 174a Lowell Observatory/National Geographic Society;175 Planet Earth Pictures; 176 NASA/SPL; 177l NASA; 177r JPL; 178 Genesis Space Photo Library; 180a Galaxy Picture Library/Carle Pieters/Brown University; 180b James Younger; 181 Russian State Archive of Science and Technology; 182 Planet Earth Pictures; 183 BBC; 184al NASA; 184ar Space Telescope Science Institute/NASA/SPL; 184bl Space Telescope Institute/NASA/SPL; 185 Galaxy Picture Library; 186al JPL; 186b United States Geological Survey/SPL; 188 NASA; 189 BBC; 190 NASA; 192 NASA/SPL; 193 Planet Earth Pictures; 194 National Geographic Society; 195 National Geographic Society; 196 Hulton Getty Picture Collection; 197 Genesis Space Photo Library; 198 Genesis Space Photo Library; 199l NASA; 199r NASA; 200l E.Imre Friedmann; 200r E.Imre Friedmann; 201a E.Imre Friedmann; 201bl Simon Fraser/SPL; 201br B.Murton/Southampton Oceanography Centre/SPL; 202 Alfred Pasieka/SPL; 203 courtesy of Ephraim Vishniac; 204 United States Geological Survey; 205 JPL; 206a J.William Schopf; 206b Simon Fraser/SPL; 207a Lowell Observatory; 207b Dr Chris Riley; 207c Lowell Observatory; 208 Steve Mojzsis; 209 Malin Space Sciences/NASA; 211 NASA/SPL; 212 NASA/SPL; 213l NASA/SPL; 213r John Frost Newspapers; 214 Space Telescope Science Institute/SPL; 216 John Sanford/SPL; 218 NASA; 222 Space Telescope Science Institute/NASA/SPL; 223a Galaxy Picture Library; 223b AKG London; 225 California Institute of Technology Palomar Observatory/Royal Observatory Edinburgh; 226 NASA; 227 Geoff Bryden; 228 Dr Fred Espenak/SPL; 229 NASA; 231 NASA

Index

Numbers in italics refer to photographs